Epitaxial Graphene on Silicon Carbide

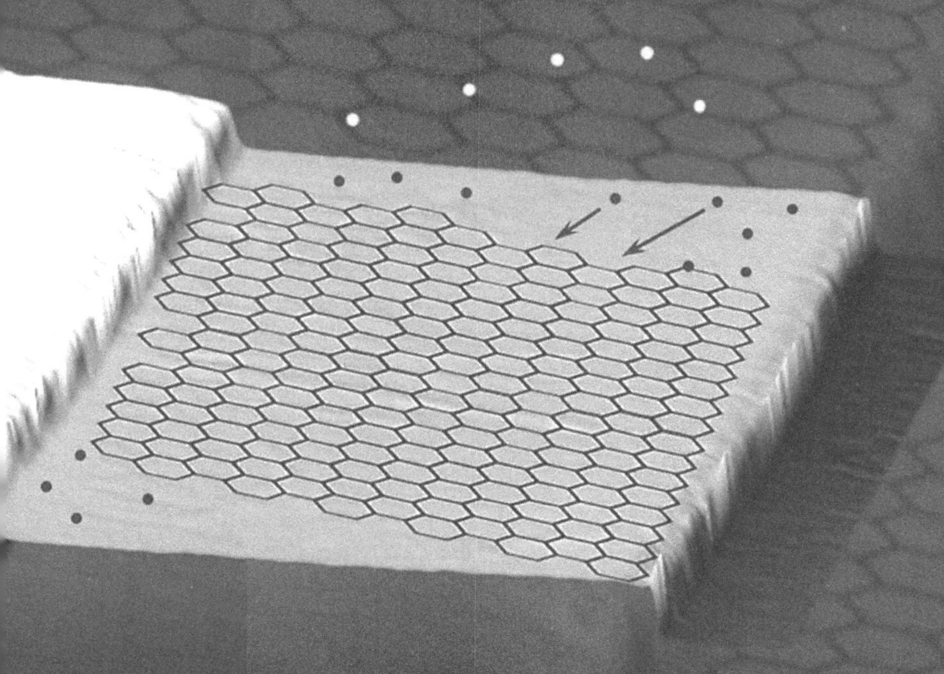

Epitaxial Graphene on Silicon Carbide

Modeling, Characterization, and Applications

edited by

Gemma Rius

Philippe Godignon

Published by

Pan Stanford Publishing Pte. Ltd.
Penthouse Level, Suntec Tower 3
8 Temasek Boulevard
Singapore 038988

Email: editorial@panstanford.com
Web: www.panstanford.com

British Library Cataloguing-in-Publication Data
A catalogue record for this book is available from the British Library.

Epitaxial Graphene on Silicon Carbide: Modeling, Devices, and Applications

ISBN 978-981-4774-20-8 (Hardcover)
ISBN 978-1-315-18614-6 (eBook)

Contents

Preface

We are pleased to present the first book dedicated exclusively to epitaxial graphene on silicon carbide (SiC). This book comprehensively addresses all relevant aspects of the study and technological development of epitaxial graphene materials and their applications. It comprises chapters contributed by a selected group of experts from Europe and Asia who have placed their research on the specific characteristics of epitaxial graphene on SiC herein in an organized manner. Importantly, each chapter provides a vision on the current state of the art of the discussed materials.

The book covers fundamental aspects of epitaxial graphene on SiC, as addressed by quantum Hall effect studies along with relevant examples on the synthesis, and probe-based characterization techniques, such as scanning tunneling microscopy and atomic resolution imaging based on transmission electron microscopy. Additionally, it addresses the processing of epitaxial graphene on SiC materials by describing pertinent methods for the fabrication of electronic devices and the fundamental aspects of their performance. It is noteworthy that it complements experimental works with theoretical modeling and simulation studies. Because of this compilation, it makes basic comprehension of epitaxial graphene on SiC substrates and its potential for electronic applications accessible.

Particularly, the book includes the state of the art on the synthesis of epitaxial graphene on SiC. It profusely explains the production process as a function of SiC substrate characteristics, such as polytype, polarity, and wafer cut, and complements the processing techniques, such as in situ and ex situ conditioning, including H_2 pre-deposition annealing, chemical–mechanical polishing, among others. It also generously describes growth studies including the most popular techniques for high-quality controlled deposition, such as ultrahigh vacuum processing and partial pressure or graphite cap–controlled sublimation techniques.

The book has been made possible by a supportive group of experts who have multidisciplinary backgrounds in physics, electronic engineering, materials science, and nanotechnology.

We sincerely acknowledge their effort and collaboration and hope that the readers will find their contributions useful and highly enlightening for learning or reviewing the main aspects of epitaxial graphene on SiC research.

Gemma Rius
Philippe Godignon
Autumn 2017

Chapter 1

Epitaxial Graphene on SiC Substrate: A View from a Specialist of SiC Growth and Materials Science

Gabriel Ferro

Laboratoire des Multimatériaux et Interfaces, UMR 5615, Université de Lyon, Université Claude Bernard Lyon1, 43 Bd du 11 Novembre 1918, 69622 Villeurbanne, France

gabriel.ferro@univ-lyon1.fr

Carbon materials (materials containing only carbon element) and silicon carbide (SiC) have always been linked. On one hand, it is obvious that SiC would not exist without graphite; on the other hand, SiC is now used for fabricating new allotropic forms of carbon. But the understanding and mastering of all these phenomena took long, mostly because SiC technology is difficult and requires reaching high temperatures. Also, for going down to the "nano world," one first needed the development of new characterization tools for identifying the new carbon objects. This led to amazing discoveries and development in the last decades. The present book represents the state of the art on the particular and promising case

Epitaxial Graphene on Silicon Carbide: Modeling, Devices, and Applications
Edited by Gemma Rius and Philippe Godignon

ISBN 978-981-4774-20-8 (Hardcover), 978-1-315-18614-6 (eBook)
www.panstanford.com

of epitaxial graphene on SiC substrate. This is the result of several years of materials science research, which implied step-by-step improvement and understanding.

The main goal of the chapter is to help the readers catch the various aspects of the materials science hiding behind the epitaxial growth of graphene on SiC. Most of the information given here can be found in other scientific books or reviews, but separately so that it is worth gathering them in the same book. In addition, the parallel between SiC and carbon materials may also provide a different and (hopefully) fruitful insight on this topic so that it is meant to be interesting for both newcomers and specialists in the field.

1.1 A Brief History of Carbon and SiC Materials

Carbon and SiC are neighboring materials, and both have found important industrial uses due to other properties than the semiconducting ones. For instance, SiC was (and is still) used in abrasive applications due to its high hardness. Graphite uses are as ancient as humankind history, e.g., drawing, heating, molten metal holding, and more recently as solid lubricant or battery anode. Diamond, the other main carbon allotropic form, was for long restricted to jewelry making. But with the discovery of semiconductors and the subsequent electronic boom in the 20th century, SiC and carbon got renewed interest. Though both materials underwent different development routes and timing, as will be detailed in the following sections, they were somehow always linked.

1.1.1 History of Carbon Materials

Today, graphene is by far the most studied carbon material due to its amazing electronic properties. But graphene is in fact among the youngest brothers of the carbon family. The oldest ones are, of course, graphite and diamond, and these two forms were the only carbon materials known for long.

1.1.1.1 Graphite and diamond

Due to their natural occurrence, graphite and diamond are known since the early days of humankind. But by simply looking at them, one would not guess that they are made of the same element. Diamond is transparent and is the hardest material on earth, while graphite is black and very brittle. These differences come only from the lattice organization of the carbon atoms (see Fig. 1.1). In graphite, each C atom is sp^2 bonded to three neighbors, forming flat sheets of hexagons, which are stacked but not covalently bonded one to each other. In diamond, C atoms are sp^3 bonded to four neighbors leading to a cubic structure alike the zinc blende one.

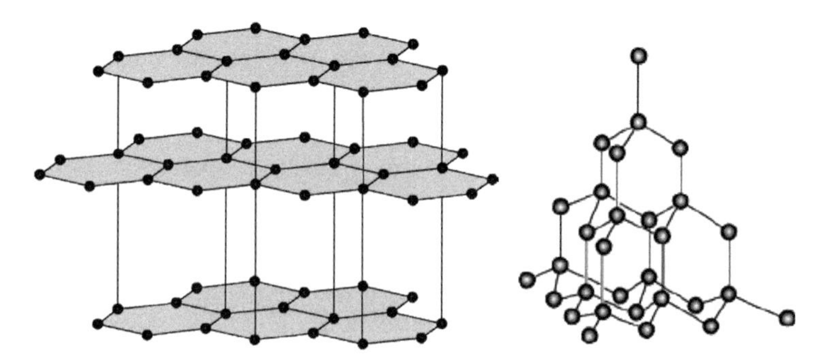

Figure 1.1 Crystalline structures of graphite (left) and diamond (right).

Graphite is the most stable allotropic form of carbon at ambient temperature and pressure, while diamond's domain of stability is at combined high temperature (>1500°C) and high pressure (> several GPa). That is why graphite is more abundant on earth than diamond. But surprisingly, it is much more difficult to obtain graphite bulk single crystals than diamond ones. On the other hand, graphite can be easily shaped in fibers, felts, or foams, while diamond cannot.

Despite the fact that these two carbon materials are very old, they are still very topical as can be seen from the fact that they are still largely used in industry and from the continuous interest of the researchers for these materials. For diamond (or sp^3 materials), the present research is mainly focused on its electronic and opto-electronic properties, while for graphite (or sp^2 materials), it obviously concerns nano-sized objects (which will be later called "new carbon materials") among which graphene is of primary importance. But before graphene, researchers have struggled a lot with other nano-sized sp^2 carbon object.

1.1.1.2 The carbon revolution

In 1985 started what can be called the "carbon revolution" with the first experimental evidence of stable C_{60} molecule (Fig. 1.2), also called Buckminsterfullerene in homage to Buckminster Fuller, whose geodesic domes it resembles [1]. This object was the first of the fullerene family, which was found later to include also C_{70}, C_{72}, C_{76}, C_{84}, and even up to C_{100} molecules. Fullerenes are usually elaborated in out-of-equilibrium conditions by vaporizing a graphite source using a laser or an electric arc discharge. Such processes are not easy to control, and they generally create many other carbon nano-objects than just fullerenes. And this is among these nano byproducts that the next carbon material, i.e., carbon nanotube (CNT), was discovered in 1991 [2]. Note that though CNT is considered part of the fullerene family (strictly speaking), I will differentiate them here for simplification purpose by considering that fullerenes are closed (and thus 0D objects).

The scientific interest for this peculiar 1D object grew rather slowly in the mid-1990s and exploded in the early 2000s, while interest for fullerene kept rather constant and moderate (Fig. 1.2). This is most probably due to the fact that CNTs can be grown selectively and in high density over a surface using a simple setup based on catalyzed chemical vapor deposition (CVD), while fullerene yield production is very low. CNT was thus a much more accessible object of study for researchers than fullerene, and the foreseen applications were more convincing. This is probably the reason why CNTs were not just a research fashion and became a real topic of interest, which is still growing as can be seen from Fig. 1.2. Note that the success of CNTs has been at the origin of the intensive research in the 2000 decade on other 1D objects, i.e., nanowires, nanoribbons, or even nanotubes made of other materials than carbon.

Then came graphene, which is often seen as the 2D version of CNTs since a graphene sheet can be theoretically obtained by unwrapping a CNT. But, contrarily to fullerenes and CNTs, at room temperature and a very simple technique, i.e., exfoliation of a piece of graphite with an adhesive tape, was used to produce graphene sheet and evidence its impressive physical properties in 2004 [3, 4]. The impact of this new carbon material on the scientific community was much stronger and more rapid than the previous ones, as can be seen from Fig. 1.2. As an illustration of such impact, the Nobel Prize in Physics was given to Geim and Novoselov only 6 years after their pioneering research on graphene. This scientific enthusiasm for graphene is obviously due to the extraordinary properties of this material, which will be described in details inside this book. Note that some more complex graphene structures (names s-, d-, and e-graphene) are also predicted, in which the base structure is not a regular honeycomb network [5]. They theoretically display different electronic properties compared to graphene and may be interesting to focus on after finding a way to produce them.

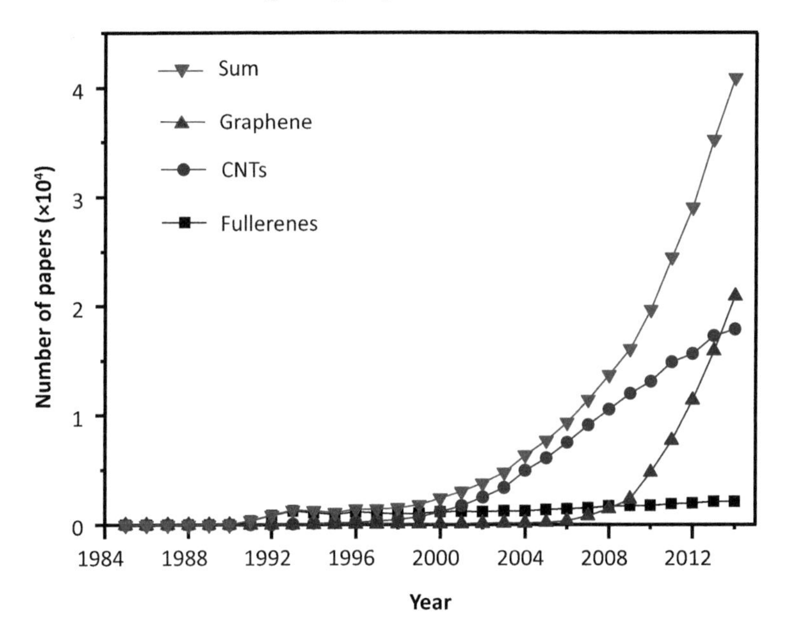

Figure 1.2 Scientific bibliometry from 1985 to 2014 using the keywords "Fullerene" (■), "CNT" (●), "Graphene" (▲) and the sum for all these keywords (▼).

Interestingly, the trend of the historical discoveries of these new carbon materials went from 0D to 1D and then 2D (see Fig. 1.3) and not the reverse as one would expect from classical miniaturization evolution. In fact, this trend was also very similar to many other semiconducting compounds for which quantum dots (0D) were studied in the 1990s, nanowires (1D) in the 2000s, and the graphene-like monolayer versions (2D) are now under investigation. It would probably be too simple to say that carbon created the fashions followed by other materials, especially for the 0D version because atomic clusters were known since decades. But this is close to the reality for 1D and 2D materials. Carbon is thus definitively a material to follow.

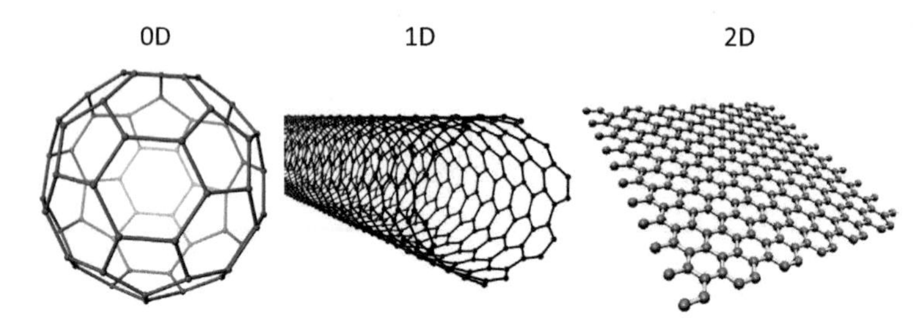

Figure 1.3 Main allotropes of carbon nanomaterials: fullerene (0D), CNT (1D), and graphene (2D).

On the structural point of view, it is important to mention that, while graphene is theoretically only composed of hexagons, fullerenes contain pentagons and heptagons, while CNTs can also contain these non-hexagonal cells as defects or at the end of a tube.

1.1.1.3 What after graphene?

Is graphene the last pure carbon material to be discovered and/or explored? This is not sure. First, carbon onions, which are spherical closed carbon shells having concentric layered structure resembling that of an onion, have been known since 1980 [6]. These peculiar objects were more marginally studied than the previously cited allotropes, but they recently attracted attention for battery and supercapacitor applications, in addition to other potential uses like in tribology or biomedical [7]. Synthesis techniques for producing carbon onions in gram-scale quantities are available so that this should not limit its industrial potentialities.

But the real next curiosities made only of carbon may be carbyne and graphyne (see Fig. 1.4). Carbyne is a chain of *sp*-hybridized carbon atoms that are linked either by alternate triple and single bonds or by consecutive double bonds. This 1D material has been predicted to be much stronger and stiffer than any other materials, but an easy or effective elaboration technique is still to be found [8, 9]. Graphyne is a 2D material analogous to graphene except that it contains both *sp*- and *sp*^{2}-hybridized C atoms [10, 11]. It could be seen as a network of C6 (benzene) units linked by carbyne bricks, though many other configurations could be imagined. But graphyne is, at the present time, purely theoretical and its properties may differ significantly from graphene ones.

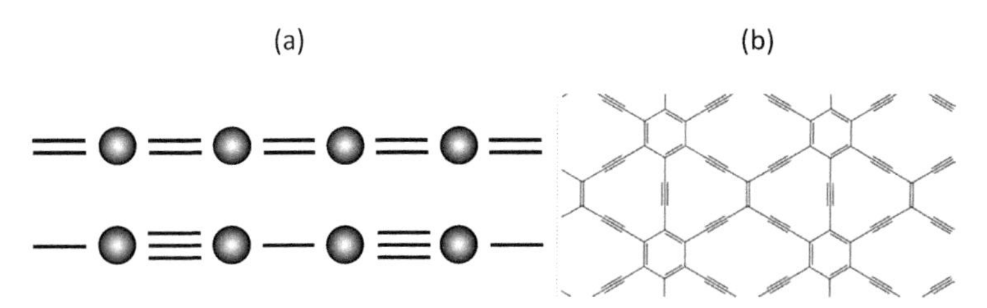

Figure 1.4 Theoretically predicted new carbon nanomaterials: (a) two possible arrangements of carbyne and (b) one (among the many) example of graphyne.

1.1.2 History of SiC Material

At the scale of human kind, SiC is a recent material compared to graphite and diamond, since it was first synthetically produced in the 19th century. This is due to the fact that SiC is very unlikely to be found in nature, except in very specific environments such as volcanic chimneys or meteorites. But since its discovery, it has been rapidly applied in industry for its unique mechanical properties. Lately came the interest for its semiconducting properties with the related difficulties in growing high crystalline quality and pure material. As can be seen from Fig. 1.5, the scientific interest for SiC material kept rather constant during the first half of the 20th century. After that, the stepped increase in scientific reports on this material was following the major technological progresses in its bulk elaboration techniques, as will be discussed in the following paragraphs.

1.1.2.1 Discovery and first industrial production

The first synthetic SiC was produced in 1824 by the Swedish chemist Jöns Jacob Berzelius in an attempt to synthesize diamond [12]. He speculated that there was a chemical bond between Si and C in one of his samples. Several decades later, in 1891, Edward Goodrich Acheson adopted an electric smelting furnace, invented in 1885 by

Eugene and Alfred Cowles, to produce suitable SiC minerals [13]. This was originally a failed experiment to produce artificial diamond out of a mixture of aluminum silicate and graphite. That is why Acheson originally named the produced blue crystals "carborundum" in reference to carbon and corundum (crystalline form of alumina) since he believed that the synthesized material was a mixture of diamond and corundum. And indeed the produced material was very hard, refractory, and infusible, almost alike the two other ones. Later, Acheson found that the crystals were made of carbon and silicon only; therefore, it could be called silicon carbide. But the company he founded for producing and selling SiC for abrasive and cutting applications kept the name carborundum. Interestingly, this inappropriate name is still used now for referring to SiC-based abrasive material.

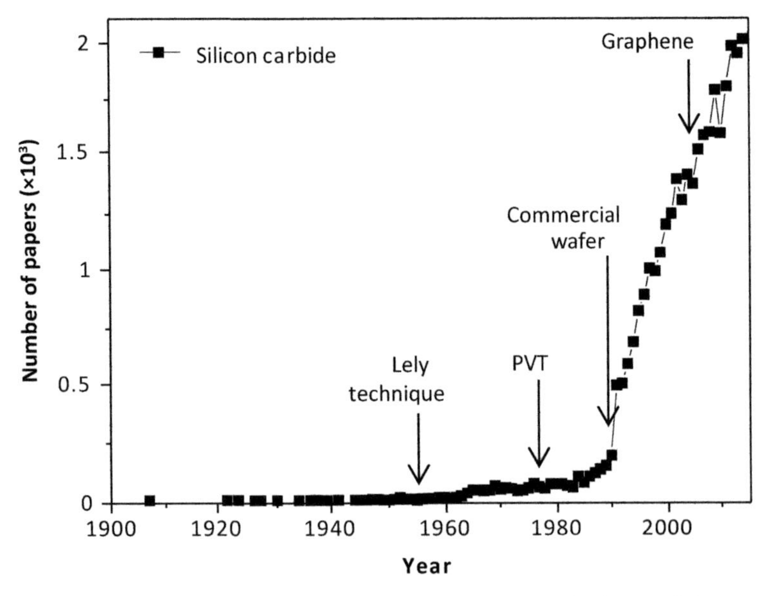

Figure 1.5 Scientific bibliometry from 1900 to 2014 using the keyword "Silicon carbide."

Such applications were for long the main ones for SiC material and they are still representing an important part of its actual mass production. In addition to its recent semiconductor related uses, SiC is also used in high-temperature, harsh-environment (nuclear, space, etc.), and high-strength applications. Except for the semiconductor applications, the nature of SiC polytype does not have a significant impact. It mainly depends on the elaboration conditions that are directly related to the end uses.

1.1.2.2 Single crystal growth

The first evidence of SiC semiconductor-like properties was found by G. W. Pierce in 1907 [14], while the same year, H. J. Round described electroluminescence on SiC crystals for the first time [15]. But the scientific interest for SiC semiconductor became really significant in the 1950s after the development of a furnace allowing the growth of crystalline platelets by J. A. Lely [16]. In the Lely process, SiC sublimes in a hot part and crystallizes in a cooler region of the furnace (Fig. 1.6a). This method was further refined by Hamilton et al. [17] and Novikov et al. [18]. It is commonly referred to as the Lely method, which produces the so-called "Lely platelets," which are cm^2-size hexagonal and flat single crystals. Starting from these crystals, scientists were able to determine in details the crystallographic, optical, and electronic properties of SiC.

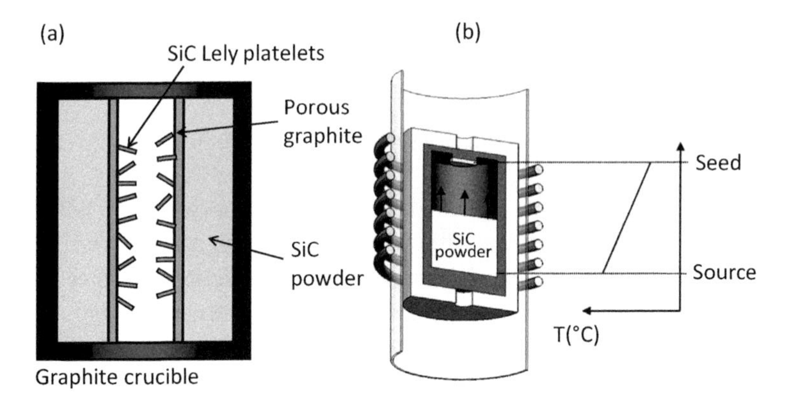

Figure 1.6 Sublimation-based techniques for producing bulk SiC crystals: (a) Lely method and (b) physical vapor transport.

A real breakthrough occurred in 1978 when Tairov and Tsvetkov [19] demonstrated the seeded growth of SiC using the sublimation method (Fig. 1.6b). It generated a renewed interest for SiC material because it could potentially produce large-area crystals. Almost all the following research works on bulk growth were only refinement and improvement of this technology. It led to commercial SiC wafers, which were first made available by Cree Research Inc., in 1991. From this date, the diameter of the commercial wafer increased regularly (Fig. 1.7) up to this year's announcement on the fabrication of 200 mm wafer by several companies.

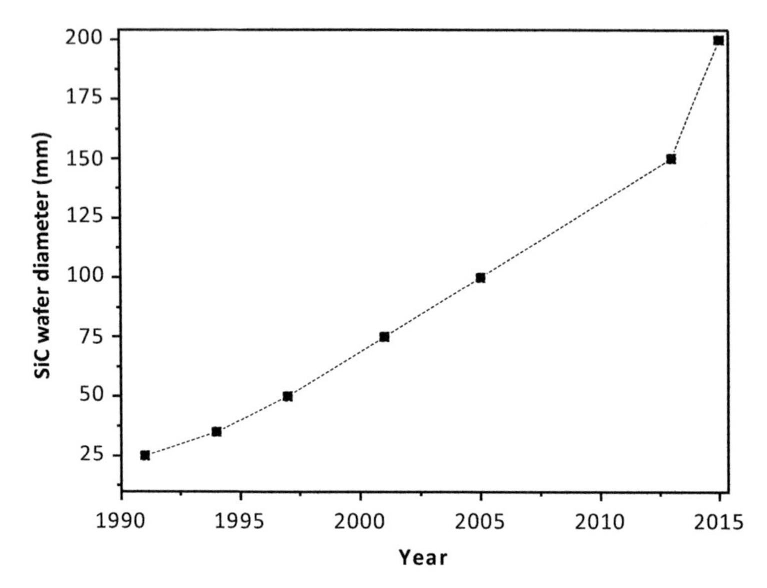

Figure 1.7 Evolution of commercial SiC wafer diameter with time.

However, it took more than 10 years between Tairov and Tsvetkov pioneer development and the first commercialization of an SiC wafer. This is due to the fact that sublimation technique, also called PVT for physical vapor transport, is very difficult to implement and control. Indeed, the PVT chamber is a black box heated at very high temperature (usually above 2200°C) and where the parameters are constantly changing during growth (graphitization of the finite SiC powder, change in the seed-powder distance and thus in the thermal gradient). We will see later that graphitization of SiC, which is seen as a drawback for SiC crystal growth, is the phenomenon at the origin of the successful growth of epitaxial graphene on SiC.

Along with the development of the SiC crystalline boules arose other technological issues to overcome in order to strengthen the viability of SiC semiconductor technology. The first ones were cutting and polishing, which were for long limiting factors because

SiC is a very hard material. Obviously, SiC surface polishing is also critical for graphene growth. But these are not the only links and inter-dependency that can be found between SiC and graphene as will be more discussed next.

1.1.2.3 Links and bridges between SiC and carbon materials

As evidenced in Fig. 1.8, SiC and carbon materials have always been historically linked. The scientific bibliometry between these materials rose up in the beginning of the 1990s similar to SiC material alone, as seen in Fig. 1.5, in relation with the commercialization of the first SiC single crystalline wafers. It is worth noting that the links between graphene and SiC started several years before the pioneer work of Geim and Novoselov, though the steep increase after 2004 is significant. But beside the obvious fact that SiC contains 50% of C atoms, what are the other links between SiC and carbon materials?

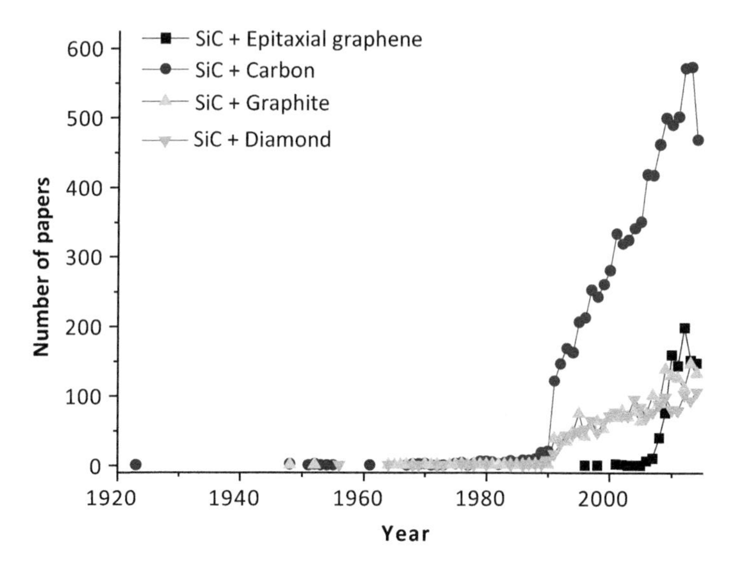

Figure 1.8 Scientific bibliometry from 1920 to 2014 using the combination of keywords "SiC and Epitaxial graphene" (■), "SiC and Carbon" (•), "SiC and Graphite" (▲), and "SiC and Diamond" (▼).

Technical links: The link between SiC and carbon materials exists since the early times of SiC discovery and production, due to the fact that graphite was the only source of C for forming SiC (see, for instance, the Acheson method). And this is still the case nowadays for most of the non-semiconductor applications of SiC. But when trying to grow bulk SiC single crystals, using an SiC source (powder) is by far the best choice. Note that in the Acheson method, graphite is not only the C source but also the heat source (resistive heating).

When considering the Lely and PVT methods for growing SiC crystals, graphite is present in large amount because providing the heat source (by RF heating in general) and being also used as the crucible material. This is due to the refractory properties of graphite, which can easily withstand the elevated temperatures required for SiC growth, in addition to the fact that it does not chemically react with SiC even at these temperatures.

For depositing epitaxial SiC layers, the current hot wall CVD reactors contain also a large amount of graphite pieces, which completely surround the substrate: bottom (susceptor), sides, and top. It increases the cracking yield of the gases and allows working at lower C excess (lower C/Si atomic ratio) in the gas phase. Note that, contrarily to PVT chambers, the graphite parts in the CVD reactors are covered with a refractory carbide coating, which can be either SiC itself or TaC. This is to avoid contamination of the SiC grown layers by the impurities usually contained in graphite (Al, B, transition metals, etc.). Note that the reactors used for CVD growth of graphene on SiC substrate are very often the same as for SiC epitaxy for two main reasons: (1) Expertise of SiC epitaxy researchers and engineers is very helpful for their knowledge on related Si and C chemistry and on proper surface preparation of samples; and (2) the SiC reactors are fully compatible with graphene growth in terms of temperature and chemistry.

Finally, composite materials are another example of the links between these C and SiC materials. For instance, SiC–C composites were intensively studied and are currently used for aeronautic and aerospace applications.

Chemical links: As mentioned before, carbon and SiC do not react when in contact, even at very high temperatures. This is mainly because SiC is the only stable compound made out of Si and C (see the Si–C phase diagram of Fig. 1.9). Such inertness is used, of course, in SiC elaboration techniques, as mentioned earlier, but it has also other uses. For instance, SiC encapsulation with graphite is often used before high-temperature (≥1600°C) annealing of implanted impurities (dopants). This is usually performed by thermal decomposition of a photoresist resin at moderate temperature before the high-temperature annealing. Applying such encapsulation allows both to reduce or suppress impurity out-diffusion and to avoid undesired/uncontrolled decomposition/graphitization of SiC at the surface.

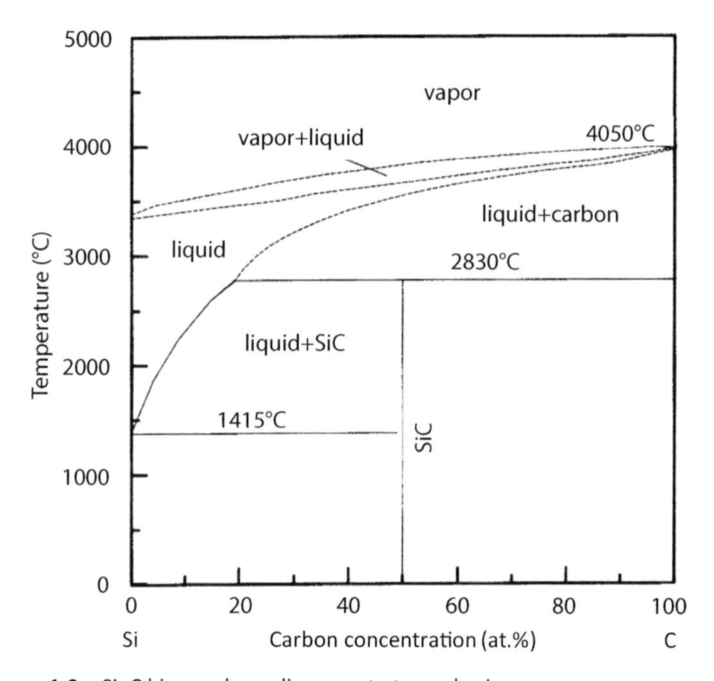

Figure 1.9 Si–C binary phase diagram at atmospheric pressure.

Such decomposition is the direct consequence of the Si–C binary phase diagram, which shows that SiC does not decompose congruently. Heating up SiC leads first to Si loss (under liquid or vapor phase) and thus C-enrichment of the solid material. Experimentally, it is known for long that when heating SiC under non-reacting atmosphere (Ar, vacuum), graphite forms on top of it. For instance, inside PVT chambers, the SiC powder (source) is at the hotter place of the crucible (see Fig. 1.6) so that it undergoes systematically Si loss and thus graphitization during the growth. When opening the crucible after such experiments, the originally green SiC powder is found black. This is among the main difficulties of the PVT process because the overall composition of the SiC source is progressively changing with time.

But one can turn it into an advantage when considering the ability to create carbon nanostructures on top of SiC by controlling this decomposition process. This is, of course, the case of epitaxial graphene formation, but one can also form several other types of carbon nanomaterials from SiC decomposition, such as CNT [20]. Note that in order to reduce the decomposition/graphitization reaction temperature, one can use the carbide-derived carbon (CDC) technique, which consists in adding halogens in the gas phase to act as Si-selective etchant. This process takes advantage of the fact that halogen-containing Si molecules are thermodynamically more stable than halogen-containing C ones and it can produce carbon onions or even nano-diamond out of SiC [21].

In conclusion, the chemical links between SiC and carbon materials are numerous and they work in both ways. Some researchers make SiC from C, while others work on making C out of SiC. Note that the latter case is mostly used for producing carbon nanomaterials, from which graphene is the most studied one.

1.2 Introduction to SiC Material

When one wants to grow an epitaxial layer on a foreign substrate, the first thing to do is to get some elementary information about this substrate. For instance, these information should help better determining the conditions window for the growth and/or the properties to be targeted for the layer itself and the final heterostructure. In the particular case of SiC substrate, there are several specificities worth knowing for best use.

1.2.1 SiC Crystallography

SiC is the only compound made out of group IV elements silicon and carbon, with the exact stoichiometry $Si_{50}C_{50}$. The Si–C bond is nearly covalent (88%), with an ionic contribution of 12% (Si positively, C negatively charged). The smallest building element of any SiC lattice is a tetrahedron defined by strong sp^3 orbitals, either SiC_4 or CSi_4 (Fig. 1.10). It means that each carbon atom is surrounded by four silicon atoms (and vice versa). Therefore, the first neighboring shell's configuration is identical for all atoms in any crystalline structure of SiC. The distance between two similar atoms, usually denoted by the letter a, is 0.308 nm. The Si–C bonding energy is 289 kJ/mol with a distance d equal to 0.189 nm. The spacing between two layers of identical atoms, which is the height of the tetrahedron, is 0.252 nm.

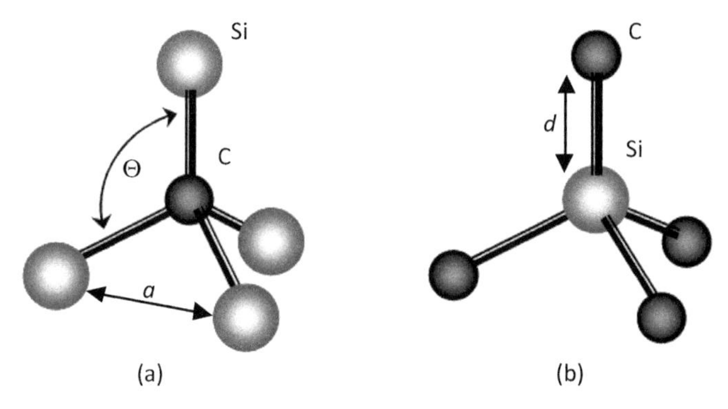

Figure 1.10 Elementary building block of SiC crystals: tetrahedrons containing (a) one C and four Si and (b) one Si and four C atoms; a = 0.308 nm, d = 0.189 nm, and Θ = 109°.

1.2.1.1 Polytypism

Silicon carbide crystals can occur in many different crystalline structures denominated as "polytypes." The phenomenon is called polytypism, which is a 1D case of polymorphism. It is rather usual to find that the number of identified SiC polytypes is higher than 200, though there are serious doubts on the validity and/or stability of many of them. Only a few of them, denoted 3C-, 4H-, 6H-SiC, and 15R, are stable enough and thus of technological interest. There is only one cubic polytype, 3C-SiC, sometimes referred to as β-SiC, while all the others are called α-SiC. The stacking sequences of the most common SiC polytypes are illustrated in Fig. 1.11.

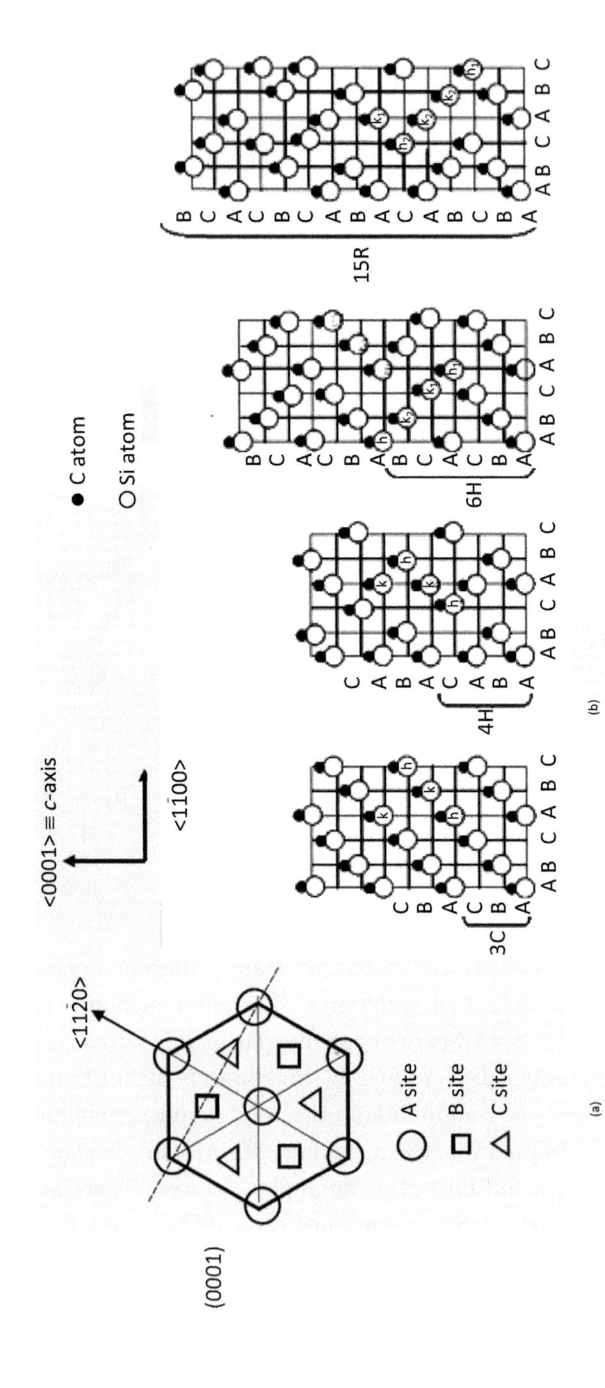

Figure 1.11 Illustration of polytypism in SiC: (a) the three possible positions (A, B, C) for Si–C bilayers stacking and (b) the stacking sequences of four common SiC polytypes along the *c*-axis. The different cubic and hexagonal lattice sites are labelled k and h, respectively.

The commonly used method to describe the SiC crystals is the one proposed by Ramsdell [22], in which a number is used according to the number of bilayers involved in the unit cell, followed by a letter that describes the crystal type (H for hexagonal, C for cubic, R for rhombohedral). It is also possible to determine the "hexagonality" of a SiC polytype, which is the percentage of the hexagonal sites out of a whole crystal. For 3C-SiC, which has only cubic sites, the hexagonality is obviously zero, whereas it is 100% for 2H-SiC (not shown here). Other polytypes have a mixing of cubic and hexagonal sites (see Fig. 1.11) so that their hexagonality varies between the two extremes (see Table 1.1 for more details).

Table 1.1 Notations, hexagonality, and other physical parameters of SiC polytypes

Ramsdell notation	ABC notation	Lattice parameters (nm) a	c	Hexa-gonality (%)	Number of non-equivalent lattice sites
3C	ABC	0.4359	0.43590	0	1
4H	ABCB	0.3073	1.0053	50	2
6H	ABCACB	0.3080	1.5117	33	3
15R	ABCACBCABACABCB	0.3079	3.7780	44	4

1.2.1.2 Polarity

Along the *c*-axis, all SiC crystals display another peculiarity, i.e., surface polarity. Indeed, an ideal single crystal of SiC cut perpendicularly to the *c*-axis on both sides will display different surface terminations: one side will be C-rich (with only C dangling bonds) and the other side will be Si-rich (with only Si dangling bonds), as illustrated in Fig. 1.12. These two faces are commonly called C-face and Si-face, with corresponding Miller index (000-1) and (0001), respectively. The direct effect of such terminations is on surface energy, which is 1.76×10^{-4} J/cm^2 and 0.71×10^{-4} J/cm^2, respectively, for Si- and C-face [23]. This directly influences the properties of SiC material, which are thus significantly different depending on surface polarity. For instance, the rates of both oxidation [24] and CVD growth [25] are faster on the C-face than on the Si-face. Impurity incorporation into SiC is also very dependent on polarity [26].

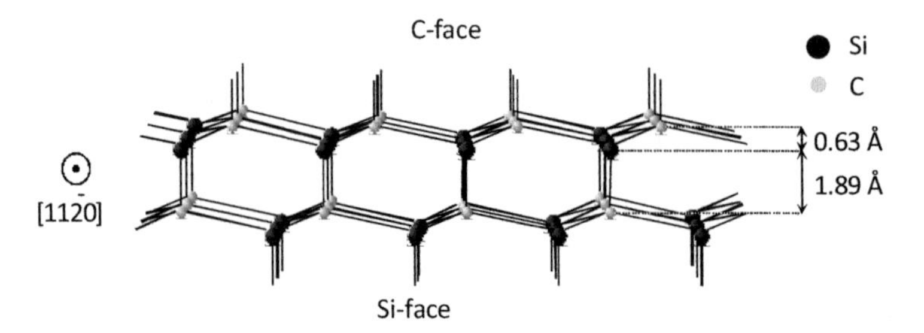

Figure 1.12 Schematic representation of the two different Si- and C-faces of SiC crystals, respectively, (0001) and (000-1) oriented.

1.2.2 Material Properties

1.2.2.1 Mechanical and chemical

Silicon carbide has traditionally been used as an abrasive material owing to its excellent mechanical properties: high hardness, high oxidation, and wear resistance (see Table 1.2 for other physical properties). SiC is also very well known for its remarkable chemical resistance to all acids and bases at room temperature. This chemical inertness can be attributed to the very strong Si–C bond, which must be broken before etching can occur. For instance, one has to heat at 400°C or higher in molten alkaline hydroxide bases, such as potassium hydroxide (KOH), in order to see a noticeable etching. Thermal etching under H_2 atmosphere is usually performed above 1400°C in order to have noticeable etching rates. If H_2 is replaced by Ar, then preferential Si loss occurs at the extreme surface leading to surface C-enrichment and graphene formation as discussed earlier. Room-temperature etching can be performed using plasma-driven techniques such as reactive ion etching (RIE) with a mixture of SF_6/O_2 reactive gases.

But when coming to reactivity toward metals, SiC does not display similar inertness. Indeed, when in contact with metals at moderate temperature (above ~400°C), it usually reacts to form stable metal carbides and/or silicides. This property, which could be detrimental for making SiC-based metal–ceramic composites, is more positively seen by electronic engineers since it allows making high-quality electric contacts to SiC electronic devices. Another important physical property of SiC is its high thermal conductivity which is almost independent on the polytype, while more dependent on microstructure or impurity incorporation. High thermal conductivity leads to reduced requirements of cooling systems, which lowers the overall system volume and cost.

Table 1.2 Mechanical properties of SiC compared to other important ceramic materials

	Hardness (kg/mm^2)	Density (g/cm^3)	Young's modulus (GPa)	Thermal conductivity (W/cmK)	Thermal expansion coefficient (/°C)
SiC	2.8×10^3	3.21	424	3.5–5	4×10^{-6}
Diamond	$8–10 \times 10^3$	3.52	1220	9–23	0.8×10^{-6}
Sapphire	$1.9–2.2 \times 10^3$	4.02	450	0.3	$3.2–5.6 \times 10^{-6}$
GaN	$1.2–1.7 \times 10^3$	6.15	320	1.3	$3–5.5 \times 10^{-6}$

Of course, in most of the mechanical applications of SiC, diamond would be a better candidate, but SiC is usually preferred because offering a good compromise between fabrication cost and performances.

1.2.2.2 Electrical and optical

It is known since long that each SiC polytype exhibits different fundamental electrical and optical properties, which are summarized in Table 1.3. Its bandgap ranges from 2.3 eV (for 3C) to 3.2 eV (for 4H), showing an almost linear increase with hexagonality [27]. Due to this large bandgap, SiC does not show any significant increase in intrinsic carrier concentration in a wide range of temperature rendering SiC devices relatively insensitive to current–voltage fluctuations even at temperature as high as 800°C, while silicon loses its semiconductor properties around 300°C. SiC's higher electron saturation drift velocity allows an increase in frequency switching capability and a decrease in power loss during such switching [28]. High-voltage electronic devices are also targeted with SiC due to its high breakdown electric field. As a consequence, this unique combination of properties makes SiC the best and the most mature candidate for high-power, high-frequency, and high-temperature electronic applications.

Table 1.3 Properties of the most stable SiC polytypes compared with other semiconducting materials

	3C-SiC	4H-SiC	6H-SiC	GaN	Diamond	Si
Energy bandgap at 300 K (eV)	2.2 (i)	3.26 (i)	3.0 (i)	3.4 (d)	5.45 (i)	1.1 (i)
Breakdown electric field ($N_D = 10^{17}/cm^3$ (MV/cm))	1.2	2	2.4	1.3	5.6	0.2
Hole mobility $N_D = 10^{16}/cm^3$ ($cm^2/V\,s$)	40	115	50	150	2000	600
Electron mobility $N_A = 10^{16}/cm^3$ ($cm^2/V\,s$)	900	1000	600	500	1900	1400
Saturation electron drift velocity (10^7 cm/s)	2.5	2.2	1.9	1.5	2.7	1
Relative dielectric constant	9.7	10	9.6	8.9	5.7	11.8
Refractive index	2.55	2.6	2.58	2.4	2.41	3.44

(i) = indirect; (d) = direct

The most common dopant impurities of SiC are Al and B for p-type and N and P for n-type, Al and N being the most used ones. These impurities in SiC affect not only the electronic properties of the material but also the optical ones. For instance, it is very easy to distinguish the polytype of n-type doped SiC bulk material just by looking to the color of the crystals (see Table 1.4). These colors are attributed to optical transitions from the lowest conduction band to other sites of increased density of states in the higher, empty bands [29].

Table 1.4 Apparent color of the SiC crystals, for the main SiC polytypes, as a function of doping type and level

	Optical bandgap at 300 K (eV)	n-type doped		p-type doped	
		n–	n+	p–	p+
3C	2.3	Yellow	Yellow–green	Dark yellow	Black
6H	3.0	Light green	Dark green	Light blue	Dark blue
4H	3.23	Light brown	Dark brown	Light blue	Dark blue

1.2.2.3 Surface preparation and thin film epitaxy

Bulk SiC substrates grown with seeded sublimation techniques cannot be used directly for the fabrication of power electronic devices. One needs to deposit epitaxially thin films on top of it with accurate control of thickness, conductivity, and dopant concentration, which values depend on the final devices targeted.

Surface preparation: As seen above, SiC is among the hardest materials known. This property was seen for long as a clear advantage for many applications, but it suddenly became a drawback when the SiC electronic era started. Indeed, in order to obtain the highest quality epitaxial layers, the seed surface must be as smooth as possible before the growth. Any imperfection at or crossing the surface (scratches, crystalline defects, roughness) can lead to defects generation inside to layer and even foreign polytype inclusions. Usually, a commercial single crystalline wafer (of any compound) has at least one of its faces mirror-like polished. Though the polishing step for "soft materials" is not too challenging, it gets very difficult with hard materials. In the case of SiC, scientists working on epitaxy during the 1990s were dealing with terrible wafer surfaces so that each group was developing its own surface preparation recipe before growth. Most of the time, this step was performed in situ in the CVD reactor and consisted in thermal etching (at temperature usually above 1500°C) under pure H_2 (which is also the carrier gas) [30] or a mixture of H_2 and HCl [31] or propane [32]. H_2 allows here to etch efficiently C atoms, while Si excess is just evaporated out of the surface. If the etching is too long or performed at too low a temperature, some Si may accumulate on the surface and form unwanted droplets. Note that this etching step has also the side benefit of removing the native oxide on the wafers. In the following decade, despite the fact that SiC surface polishing was improving significantly, epitaxy scientists kept using their surface preparation step though adapting it to the new generations of reactors, to the wafer size increase, and to reduce the density of basal plane dislocations inside the grown layer [33].

Note that SiC polishing can still be considered an industrial secret, and companies selling SiC wafers are very reluctant giving details on their polishing procedures.

SiC epitaxial growth: Though several techniques have been investigated for the epitaxial growth of SiC, CVD has proven to be the only one able to reach the targeted specifications of quality, uniformity, and reproducibility for devices mass production. As a matter of fact, CVD is the current research and industrial standard for SiC epitaxial growth. But this did not happen suddenly, and the technique benefited from several decades of improvement. Indeed the history of SiC deposition by CVD dates back to the 1960s with chlorinated precursors ($SiCl_4$ + CCl_4 [34] or CH_3SiCl_3 [35]). Later came the use of alcane- and/or silane-based precursor mixture, but the deposit was hardly monocrystalline (except for whisker growth). In the 1980s, the successful demonstration by Nishino et al. [36, 37] of 3C-SiC heteroepitaxy on silicon substrate using silane and propane promoted a renewed interest in the field of SiC epitaxy.

In this pioneering work, a cold wall CVD reactor used for GaAs epitaxy was redesigned to meet the requirements for SiC growth. Since then a lot of efforts have been made to improve the reactor design, growth processes, and reactor capacity to fulfill the requirements on the quality of the grown layers and to lower the overall cost of the process. One of the major evolutions came from the use of a hot wall configuration from Janzén's group in 1995 [38], which allowed increasing precursor cracking efficiency, growth rate, and layers purity, while improving heating power efficiency. This became the standard with some further developments (substrate rotation, low pressure, multi-wafer, vertical/horizontal reactor, chlorinated precursors).

On the scientific aspect, a major advance was brought by Matsunami's group, who introduced the concept of "step-controlled epitaxy," which allowed reducing the growth of homoepitaxial α-SiC layers down to 1400–1500°C [39]. From this time, (0001) off-axis has been the wafers standard for homoepitaxial growth. A second important step was the demonstration of dopant incorporation control via the C–Si ratio in the gas phase, called also "site competition" doping [40], which gave an easy way of monitoring the purity and/or intentional doping level of epitaxial layers. Then came in the 2000s the race for high growth rates for fabricating high-voltage devices, for which chlorinated precursors took an important place. Record values above 100 μm/h were demonstrated by several groups but not used industrially yet.

Even though SiC growth conditions are reactor dependent, their typical working ranges are $1500 \leq T \leq 1650$°C, $P \leq 100$ mbar, $0.5 \leq$ C–Si ratio ≤ 3, and $5 \leq G_R \leq 50$ μm/h. Layers purity can go below 10^{14}/cm^3 and doping level (both n- and p-type) can be controlled up to 10^{19}–10^{20}/cm^3 range.

1.3 Methods for Growth of Epitaxial Graphene on SiC

When one uses the term of epitaxy, this refer to a crystalline growth over a crystalline substrate, with inheritance of the substrate orientation inside the layer. In the case of epitaxial graphene (EG), this is a bit more complex because this material has a very specific structure (see Fig. 1.3), which is not found among the commercially available crystalline substrates. So the epitaxial relationship between graphene and the underlying substrate is often not trivial [41]. Despite this, graphene can order itself on SiC but also on many different kinds of substrates such as Cu, Ni, Ir, and Ru [42]. This is because the EG has a degree of freedom that is usually not found in other epitaxial systems: Ideally, all the C atoms are bonded together with no out-of-graphene sheet bonds so that the graphene sheets are theoretically not covalently bonded to the substrates and can thus move on it. That is why graphene easily forms pleats due to the difference in thermal expansion with the substrate [43].

At the beginning of EG research, most of the publications were mentioning the use of SiC substrate, but this is not anymore true. As can be seen in Fig. 1.13, the number of papers dealing with EG and SiC is today almost the same as EG on other substrates. But, strictly speaking, the term "epitaxial" for a graphene sheet deposited on a polycrystalline metallic surface is obviously abusively used and should only concern the use of crystalline substrates, i.e., mostly SiC. As a proof, in the case of EG grown on 3C-SiC(100)/Si(100) template, the in-plane orientation of the graphene sheets can rotate 90° to follow the one of the underlying SiC materials, which contains anti-phase domains (APD) [44]. Another specific feature of SiC is the fact that this is the only large-area C-containing substrate, which allows EG formation without the need of any reactive gas, i.e., by simple high-temperature graphitization under Ar or vacuum.

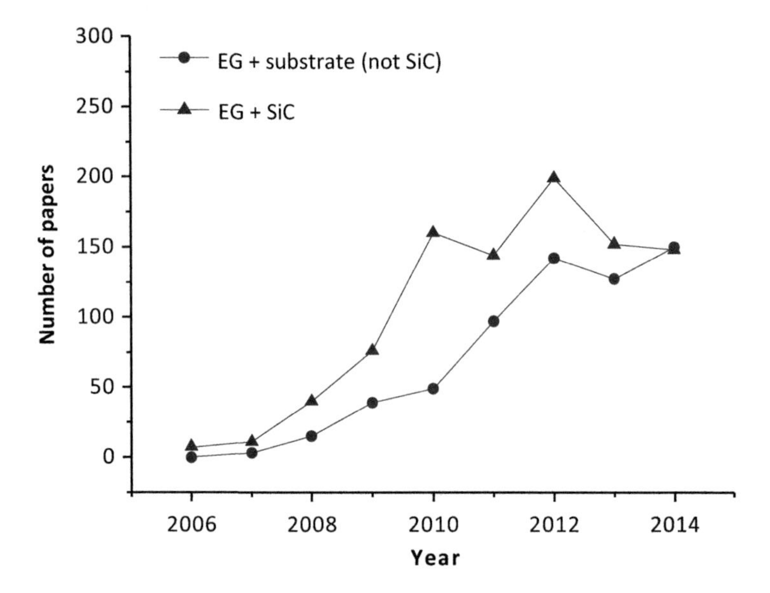

Figure 1.13 Scientific bibliometry using keywords combination "Epitaxial graphene + SiC" (▲) and "Epitaxial graphene + substrate – SiC" (•).

Before detailing the methods for growing EG on SiC, it is worth mentioning that several types of SiC substrates are available and, depending on the type of SiC substrate, growth conditions may vary significantly. This is summarized in Table 1.5.

Since the two main methods for growing EG on SiC will be largely detailed in the other chapters, we will present here only a short introduction and we invite the readers to go to Chapters 2–4 for more and specific information.

Table 1.5 Different types of SiC substrates on which EG has been grown and respective relevant points to take into account

Type of SiC substrate	Advantages	Drawbacks
α-SiC (0001): Si-face, On-axis	Commercial, large area, and very flat Layer-by-layer growth: easy control of single to few layers EG	Expensive Not fully on-axis Formation of a transition layer that acts as dopant to the EG
α-SiC (0001): Si-face, Off-axis	Commercial, large area, and very flat Facet-induced EG growth No transition layer on facets	Expensive Faster growth on facets Small terraces Growth on both on-axis planes and facets
α-SiC (000-1): C-face, On-axis and off-axis	Commercial, large area, and very flat No transition layer	Expensive 3D-like growth Difficulty to obtain large-area single-layer EG
3C-SiC(111) on Si(111)	Commercial, large area Cheaper than α-SiC	Average crystalline quality and rough surface Temperature limited by Si melting point
3C-SiC(100) on Si(100)	Commercial, large area Cheaper than 4H-SiC	Average crystalline quality and rough surface Temperature limited by Si melting point APD-induced rotation of EG
3C-SiC(111) on α-SiC (0001) Si-face	Temperature not limited by Si melting point Possibility of rhombohedral EG growth [45]	Not commercially available Small size

1.3.1 SiC Graphitization

As discussed earlier, when heating SiC to a sufficiently high temperature under inert ambience, it starts to undergo a non-congruent decomposition by losing first the Si atoms from its surface. The kinetics of Si loss depends mainly on two parameters: temperature and pressure. As a general trend for EG formation, using lower pressure allows reducing the optimal temperature. This is simply following Le Chatelier's principle, as Si loss (vaporization) results from SiC's reaction toward pressure lowering. But one needs to heat up substantially SiC in order to start breaking the Si–C bonds. As a matter of fact, just by tuning the pressure, one can grow EG on SiC from 1250°C in ultrahigh vacuum (UHV) [46] up to 2000°C under 1 atm of Ar [47]. Note that the EG formation temperature may be further reduced down to 1150°C by N_2 plasma pre-treatment [48] or even to 1100°C by adding Ni–Cu catalytic alloy on SiC surface [49] before annealing.

The growth step is often performed after an in situ surface preparation for removing the native oxide and/or any other surface contaminants. This preparation step can be performed either under pure H_2, mixture of Ar and H_2, or even under Si flux for UHV cases [46]. The addition of Si atoms in the gas phase is an additional parameter that can allow further control of the graphitization process by partially counterbalancing the Si loss and thus slowing down the EG growth kinetics [50]. Note that N_2 is not an inert gas for EG formation. Indeed, it is an n-type dopant of graphene and it can form C_xN_y volatile species in contact with C atoms at elevated temperatures. But N_2 may be used for in situ doping of EG during its growth [51].

1.3.2 CVD of EG on SiC

Another significant way of EG growth on SiC is to bring the C atoms from the gas phase just like the growth on metallic substrates. But the comparison stops here because, in the case of SiC substrate, there is no catalytic effect or any C dissolution/precipitation process. As a matter of fact, using standard thermal CVD, the parameters that can be tuned for EG growth on SiC are rather classical: precursor type and concentration, temperature and pressure. Concerning the C precursor, propane is the most used one [52–54], which is probably an inheritance of the SiC epitaxial growth chemistry since the SiC CVD reactors are often used for graphene, as discussed before. But, unlike SiC epitaxy, Ar vector gas for EG CVD is preferentially used instead of H_2. This is due to the H_2 reducing/etching effect at the deposition temperature, which is less favorable for C deposit formation. Note that this effect can be counterbalanced by increasing propane partial pressure [52, 55]. Using H_2 ambience can also lead to H intercalation below the EG and thus disappearance of the transition layer on the Si face [55].

In the absence of surface catalytic effect, the deposition temperature range is limited by the precursor cracking so that depositions are usually performed above 1300°C. And due to these high temperatures, the SiC substrate may undergo some Si loss (when performed under Ar atmosphere) before or even during EG growth so that the overall mechanism is probably a mixing of SiC decomposition and pure CVD. Note that significant temperature lowering down to 540°C can be achieved by using plasma assistance for the cracking [56]. In that case, even a more stable precursor such as CH_4 can be used.

1.4 Challenges and Perspectives

This section will only focus on the material aspects of the challenges and perspectives of EG graphene on SiC substrate, i.e., mostly on growth-related aspects.

1.4.1 Graphene Growth and Doping Optimization

Despite the considerable international effort on EG growth on SiC substrate, there is still a lot of research to do for a better understanding of the processes and for optimizing the material itself. One of the interests for using SiC for EG growth is that high crystalline quality and very flat 4H-SiC wafers are readily available in rather large areas, up to 150 mm diameter now and soon will come 200 mm crystals. But are the actual machines, designed for SiC epitaxy, capable of handling such large wafers while keeping good uniformity over the wafers, and from wafer to wafer? This is a question that has to be answered knowing that the term uniformity includes several parameters such as thickness, doping, and/or electronic properties. The present efforts on large-area SiC epitaxy will most probably help in going in the right direction for graphene.

Growth on structured SiC surface is a recent trend, which provides another degree of freedom not only to the growers but also for improvement in electronic properties [57–59]. This can lead, for instance, to quasi-freestanding graphene nanoribbons [60] or to anisotropy of the electronic properties [61]. Again, expertise from SiC material specialists is of great help to find smart solutions for patterning or self-structuring of the substrate surface.

Doping control of EG is also another important issue under current investigation. Both doping types, n and p, are of interest. For n-type, N incorporation can be used, either during EG growth [51, 62] or after growth [63, 64]. For the p-type, it is more difficult because p dopants would be elements of the IIIb column (B, Al, Ga, ...), which are not as easy to implement as N_2 gas. Furthermore, B and Al elements are known to form stable solid carbides, Al_4C_3 and B_4C, respectively, which is not the case for N. But, if one intercalates H or F atoms at the SiC–EG interface, graphene tends to be p-type [65, 66]. This decoupling of EG from the substrate is also a topic of interest because it can improve the electronic properties of graphene, for instance with N atoms [67, 68], or even allow fabrication of nano p–n junctions [69]. But even without going up to the decoupling, some optimization of the transition layer between EG and SiC may also be worth studying [70].

Note that all the recent trends reported above can be implemented either during graphitization or CVD growth of EG on SiC, even though graphitization is still the main studied approach. So one has the right to ask the following question: Which approach gives better results and/or is more promising between CVD and graphitization? The answer is not simple because, as usual, both methods have advantages and drawbacks. The future will probably tell us the answer.

1.4.2 Other Graphene-like 2D Materials on SiC

When considering the possibility of growing other 2D graphene-like materials on SiC substrate, the first one that comes in mind is silicene, a layered material composed of sp^2-hybridized silicon. This material has already been successfully grown on Ag, Ir, or ZrB_2 surfaces [71] but not on SiC. This is somewhat surprising when just considering the reverse reaction of SiC graphitization, i.e., Si surface enrichment or silicidation by C preferential loss. But this approach is intrinsically limited by SiC thermo-chemistry itself. Indeed, C atoms are less volatile than Si ones upon heating so that one should need some reaction-assisted C loss. It means finding a reactant that selectively etches away C atoms without forming any compound with Si. Such a reactant is still to be found. Additionally, in the case of Si, sp^3-hybridized state is more stable than the sp^2-hybridized one so that the inevitable formation of chemical bonding with the SiC substrate should promote sp^3 hybridization instead of the sp^2 required for ideal silicene 2D material. As a consequence, if a solution for growing silicene on SiC is to be found, it may be most probably using an Si deposition approach rather than an SiC decomposition one.

Boron nitride (BN) is another obvious 2D material (called boronitrene in its 2D form) that can be considered because of its chemical inertness with SiC and its crystallographic analogy with graphene [71]. Despite such advantages, very few works on boronitrene formation on SiC substrate can be found [72], though a bit more are available for thicker BN material [73, 74]. This is because BN epitaxy is difficult and having an intermediate graphene-like layer helps improving the layer quality [75]. But increasing thickness degrades the material [73]. Things may be a bit simpler for boronitrene (few-layer BN only) as shown by Shin et al. who successfully deposited boronitrene by CVD on SiC [72]. After annealing the system at high temperature, the BN layer was partially degraded and replaced so that h-BN/graphene lateral structure could be obtained. Note also the demonstration of physical transfer of boronitrene sheets, initially grown on Cu foil by catalytic thermal CVD, onto EG/SiC [76].

Transition metal dichalcogenides (TMDCs) 2D sheets, such as MoS_2 or WSe_2, can also be grown on SiC [77–80]. And like for boronitrene, an EG interlayer helps to get some epitaxy (called van der Waals epitaxy), but this is not mandatory [80]. As a matter of fact, both TMDC/SiC and TMDC/EG/SiC heterostructures can be fabricated giving another degree of exploration for both growers and device makers.

1.5 Conclusion

The main goal of this introduction was to show that graphite and SiC are more than just occasional friends. They can be seen as two old cousins living in the neighborhood and meeting regularly for their mutual benefit. From their last encounter came partly out the graphene boom, which could lead to a real win–win situation in which high-quality graphene can find an ideal large-area substrate to grow on, and in which SiC can enter new markets. The potentialities are high, but the efforts to be made remain important. This will be better explained by the authors of the subsequent chapters of this book.

References

1. Kroto, H. W., Heath, J. R., O'Brien, S. C., Curl, R. F., and Smalley, R. E. (1985). C_{60}: Buckminsterfullerene, *Nature*, **318**, pp. 162–163.
2. Iijima, S. (1991). Helical microtubules of graphitic carbon, *Nature*, **354**, pp. 56–58.
3. Novoselov, K. S., Geim, A. K., Morozov, S. V., Jiang, D., Zhang, Y., Dubonos, S. V., Grigorieva, I. V., and Firsov, A. A. (2004). Electric field effect in atomically thin carbon films, *Science*, **306**(5696), pp. 666–669.
4. Novoselov, K. S., Geim, A. K., Morozov, S. V., Jiang, D., Katsnelson, M. I., Grigorieva, I. V., Dubonos, S. V., and Firsov, A. A. (2005). Two-dimensional gas of massless Dirac fermions in graphene, *Nature*, **438**, pp. 197–200.
5. Xu, L. C., Wang, R. Z., Miao, M. S., Wei, X. L., Chen, Y. P., Yan, H., Lau, W. M., Liu, L. M., and Ma, Y. M. (2014). Two-dimensional Dirac carbon allotropes from graphene, *Nanoscale*, **6**, pp. 1113–1118.
6. Iijima, S. (1980). Direct observation of the tetrahedral bonding in graphitized carbon-black by high-resolution electron-microscopy, *J. Cryst. Growth*, **50**, pp. 675–683.
7. Bartelmess, J., Giordani, S., and Beilstein, J. (2014). Carbon nano-onions (multi-layer fullerenes): Chemistry and applications, *Nanotechnology*, **5**, pp. 1980–1998.
8. Cannella, C. B. and Goldman, N. (2015). Carbyne fiber synthesis within evaporating metallic liquid carbon, *J. Phys. Chem. C*, **119**, pp. 21605–21611.

9. Bartik, T., Bartik, B., Monika, B., Dembinski, R., Gladysz, J. A. (1996). A step-growth approach to metal-capped one-dimensional carbon allotropes: Syntheses of C_{12}, C_{16}, and C_{20} μ-polyynediyl complexes, *Angew. Chem. Int. Ed. Engl.*, **35**(4), pp. 414–417.

10. Kondo, M., Nozaki, D., Tachibana, M., Yumura, T., and Yoshizawa, K. (2005). Electronic structures and band gaps of chains and sheets based on phenylacetylene units, *Chem. Phys.*, **312**, pp. 289–297.

11. Tahara, K., Yoshimura, T., Sonoda, M., Tobe, Y., and Williams, R. V. (2007). Theoretical studies on graphyne substructures: Geometry, aromaticity, and electronic properties of the multiply fused dehydrobenzo[12] annulenes, *J. Org. Chem.*, **72**, pp. 1437–1442.

12. Berzelius, J. J. (1824). Untersuchungen über die Flusspath Faure und deren merkwürdige Verbindungen, *Ann. Phys., Lpz.*, **1**, pp. 169.

13. Acheson, E. G. (1892). Production of artificial crystalline carbonaceous materials, carborundum, *Engl. Patent 17911*.

14. Pierce, G. W. (1907). Crystal rectifiers for electric currents and electric oscillations. Part I. Carborundum, *Phys. Rev. (Series I)*, **25**, pp. 31.

15. Round, H. J. (1907). A note on Carborundum, *Electrical World*, **49**, pp. 309.

16. Lely, J. A. (1955). Darstellung von einkristallen von silicium carbid und beherrschung von art und menge der eingebautem venmreingungen, *Bericht. Deutschen Keram. Ges.*, **32**, pp. 229.

17. Hamilton, D. R. (1960). The growth of silicon carbide by sublimation, in: *Silicon Carbide: A High Temperature Semiconductor*, Smilestens, J. (ed). Pergamon, Oxford, pp. 45–51.

18. Novikov, V. P. and Ionov, V. I. (1968). Production of monocrystals of alpha-silicon carbide, *Growth Crystals*, **6b**, pp. 9–21.

19. Tairov, Y. M. and Tsvertkov, V. F. (1978). Investigation of growth processes of ingots of silicon-carbide single-crystals, *J. Cryst. Growth*, **43**, pp. 209–212.

20. Kusunoki, M., Rokkaku, M., and Suzuki, T. (1997). Epitaxial carbon nanotube film self-organized by sublimation decomposition of silicon carbide, *Appl. Phys. Lett.*, **71**(18), pp. 2620–2622.

21. Welz, S., McNallan, M. J., and Gogotsi, Y. (2006). Carbon structures in silicon carbide derived carbon, *J. Mater. Processing Techn.*, **179**, pp. 11–22.

22. Ramsdell, L. S. (1947). Studies on silicon carbide, *Am. Mineral.*, **32**, pp. 64–82.

23. Syvajarvi, M., Yakimova, R., and Janzen, E. (1999). Interfacial properties in liquid phase growth of SiC, *J. Electrochem. Soc.*, **146**(4), pp. 1565–1569.

24. Harris, R. C. A. (1975). Oxidation of 6h-alpha silicon-carbide platelets, *J. Am. Ceram. Soc.*, **58**(1–2), pp. 7–9.

25. Kimoto, T. and Matsunami, H. (1994). Nucleation and step motion in chemical vapor deposition of SiC on 6H-SiC{0001} faces, *J. Appl. Phys.*, **76**(11), pp. 7322–7327.

26. Larkin, D. J. (1997). SiC dopant incorporation control using site-competition CVD, *Phys. Stat. Sol.*, **202**, pp. 305–320.

27. Haeringen, W. V., Bobbert, P. A., and Backes, W. H. (1997). On the band gap variation in SiC polytypes, *Phys. Stat. Sol.*, **202**, pp. 63–80.

28. Baliga, B. J. (2010). *Fundamentals of Power Semiconductor Devices.* Springer Science & Business Media, ISBN 978-0-387-47313-0, p. 1072.

29. Harris, G. L. (1995). Optical absorption and refractive index of SiC, in: *Properties of Silicon Carbide, EMIS Datareviews Series 13*, INSPEC, London, ISBN 0852968701.

30. Xie, Z. Y., Wei, C. H., Li, L. Y., Yu, Q. M., and Edgar, J. H. (2000). Gaseous etching of 6H-SiC at relatively low temperatures, *J. Cryst. Growth*, **217**, pp. 115–124.

31. Leone, S., Henry, A., Janzén, E., and Nishizawa, S. (2013). Epitaxial growth of SiC with chlorinated precursors on different off-angle substrates, *J. Cryst. Growth*, **362**, pp. 170–173.

32. Soueidan, M., Ferro, G., Dazord, J., Monteil, Y., and Younes, G. (2005). Surface preparation of α-SiC for the epitaxial growth of 3C–SiC, *J. Cryst. Growth*, **275**(1–2), pp. e1011–e1016.

33. Stahlbush, R. E., van Mil, B. L., Myers-Ward, R. L., Lew, K.-K., Gaskill, D. K., and Eddy Jr., C. R. (2009). Basal plane dislocation reduction in 4H-SiC epitaxy by growth interruptions, *Appl. Phys. Lett.*, **94**, 041916.

34. Jennings, V. J., Sommers, A., and Chang, H. C. (1966). Epitaxial growth of silicon carbide, *J. Electrochem. Soc.*, **113**, pp. 728–731.

35. Ivanovna, L. M. and Pletiouchkine, A. A. (1968). *Izvest. Akad. Nank. SSSR, Neorg. Mater.*, **4**(7), p. 1089.

36. Matsunami, H., Nishino, S., and Ono, H. (1981). Heteroepitaxial growth of cubic silicon carbide on foreign substrates, *IEEE Trans. Electron Devices*, **28**(10), pp. 1235–1236.

37. Nishino, S., Powell, J. A., and Will, H. A. (1983). Production of large-area single-crystal wafers of cubic SiC for semiconductor-devices, *Appl. Phys. Lett.*, **42**(5), pp. 460–462.

38. Kordina, O., Björketun, L. O., Henry, A., Hallin, C., Glass, R. C., Hultman, L., Sundgren, J. E., and Janzén, E. (1995). Growth of 3C-SiC on on-axis Si(100) substrates by chemical vapor deposition, *J. Cryst. Growth*, **154**, pp. 303–314.

39. Kuroda, N., Shibahara, K., Yoo, W. S., Nishino, S., and Matsunami, H. (1987). *Extended Abstracts 19th Int. Conf. Solid State Devices and Materials*, Tokyo, pp. 227.

40. Larkin, D. J., Neudeck, P. G., Powell, J. A., and Matus, L. G. (1994). Site-competition epitaxy for superior silicon carbide electronics, *Appl. Phys. Lett.*, **65**(13), pp. 1659–1661.

41. Riedl, C., Coletti, C., and Starke, U. (2010). Structural and electronic properties of epitaxial of epitaxial graphene on SiC(0001): A review of growth, characterization, transfer doping and hydrogen intercalation, *J. Phys. D Appl. Phys.*, **43**(37), 374009 (17pp).

42. Dedkov, Y. and Voloshina, E. (2015). Graphene growth and properties on metal substrates, *J. Phys. Condens. Matter*, **27**, 303002.

43. Hu, Y., Ruan, M., Guo, Z., Dong, R., Palmer, J., Hankinson, J., Berger, C., and De Heer, W. A. (2012). Structured epitaxial graphene: Growth and properties, *J. Phys. D Appl. Phys.*, **45**, 154010.

44. Ouerghi, A., Balan, A., Castelli, C., Picher, M., Belkhou, R., Eddrief, M., Silly, M. G., Marangolo, M., Shukla, A., and Sirotti, F. (2012). Epitaxial graphene on single domain 3C-SiC(100) thin films grown on off-axis Si(100), *Appl. Phys. Lett.*, **101**, 021603.

45. Pierucci, D., Sediri, H., Hajlaoui, M., Girard, J. C., Brumme, T., Calandra, M., Velez-Fort, E., Patriarche, G., Silly, M. G., Ferro, G., Souliere, V., Marangolo, M., Sirotti, F., Mauri, F., and Ouerghi, A. (2015). Evidence for flat bands near the fermi level in epitaxial rhombohedral multilayer graphene, *ACS Nano*, **9**(5), pp. 5432–5439.

46. Ouerghi, A., Belkhou, R., Marangolo, M., Silly, M. G., El Moussaoui, S., Eddrief, M., Largeau, L., Portail, M., and Sirotti, F. (2010). Structural coherency of epitaxial graphene on 3C-SiC(111) epilayers on Si(111), *Appl. Phys. Lett.*, **97**, 161905.

47. Virojanadara, C., Syväjarvi, M. Yakimova, R., Johansson, L. I., Zakharov, A. A., and Balasubramanian, T. (2008). Homogeneous large-area graphene layer growth on 6H-SiC(0001), *Phys. Rev. B*, **78**, 245403.

48. Tsai, H. S., Lai, C. C., Medina, H., Lin, S. M., Shih, Y. C., Chen, Y. Z., Liang, J. H., and Chueh, Y. L. (2014). Scalable graphene synthesised by plasma-assisted selective reaction on silicon carbide for device applications, *Nanoscale*, **6**, pp. 13861–13869.

49. Iacopi, F., Mishra, N., Cunning, B. V., Goding, D., Dimitrijev, S., Brock, R., Dauskardt, R. H., Wood, B., and Boeckl, J. (2015). A catalytic alloy approach for graphene on epitaxial SiC on silicon wafers, *J. Mater. Res.*, **30**(5), pp. 609–616.

50. Srivastava, N., He, G., Luxmi, Mende, P. C., Feenstra, R. M., Sun, Y. (2012). Graphene formed on SiC under various environments: Comparison of Si-face and C-face, *J. Phys. D Appl. Phys.*, **45**(15), 154001.

51. Velez-Fort, E., Mathieu, C., Pallecchi, E., Pigneur, M., Silly, M. G., Belkhou, R., Marangolo, M., Shukla, A., Sirotti, F., Ouerghi, A. (2012). Epitaxial graphene on 4H-SiC(0001) grown under nitrogen flux: Evidence of low nitrogen doping and high charge transfer, *ACS Nano*, **6**, 10893.

52. Michon, A., Vézian, S., Ouerghi, A., Zielinski, M., Chassagne, T., and Portail, M. (2010). Direct growth of few-layer graphene on 6H-SiC and 3C-SiC/Si via propane chemical vapor deposition, *Appl. Phys. Lett.*, **97**, 171909.

53. Strupinski, W., Grodecki, K., Wysmolek, A., Stepniewski, R., Szkopek, T., Gaskell, P. E., Grüneis, A., Haberer, D., Bozek, R., Krupka, J., and Baranowski, J. M. (2011). Graphene epitaxy by chemical vapor deposition on SiC, *Nano Lett.*, **11**(4), pp. 1786–1791.

54. Hwang, J., Shields, V. B., Thomas, C. I., Shivaraman, S., Hao, D., Kim, M. R., Woll, A., Tompa, G. S., and Spencer, M. G. (2010). Epitaxial growth of graphitic carbon on C-face SiC and sapphire by chemical vapor deposition (CVD), *J. Cryst. Growth*, **312**(21), pp. 3219–3224.

55. Michon, A., Vezian, S., Roudon, E., Lefebvre, D., Zielinski, M., Chassagne, T., and Portail, M. (2013). Effects of pressure, temperature, and hydrogen during graphene growth on SiC(0001) using propane-hydrogen chemical vapor deposition, *J. Appl. Phys.*, **113**, 203501.

56. Zhang, L., Shi, Z., Liu, D., Yang, R., Shi, D., and Zhang, G. (2012). Vapour-phase graphene epitaxy at low temperatures, *Nano Res.*, **5**(4), pp. 258–264.

57. Giannazzo, F., Deretzis, I., Nicotra, G., Fisichella, G., Ramasse, Q. M., Spinella, C., Roccaforte, F., and La Magna, A. (2014). High resolution study of structural and electronic properties of epitaxial graphene grown on off-axis 4H-SiC (0001), *J. Cryst. Growth*, **393**, pp. 150–155.

58. Palacio, I., Celis, A., Nair, M. N., Gloter, A., Zobelli, A., Sicot, M., Malterre, D., Nevius, M. S., De Heer, W. A., Berger, C., Conrad, E. H., Taleb-Ibrahimi, A.,

and Tejeda, A. (2015). Atomic structure of epitaxial graphene sidewall nanoribbons: Flat graphene, miniribbons, and the confinement gap, *Nano Lett.*, **15**, pp. 182–189.

59. Kruskopf, M., Pierz, K., Wundrack, S., Stosch, R., Dziomba, T., Kalmbach, C. C., Müller, A., Baringhaus, J., Tegenkamp, C., Ahlers, F. J., and Schumacher, H. W. (2015). Epitaxial graphene on SiC: Modification of structural and electron transport properties by substrate pretreatment, *J. Phys. Condens. Matter*, **27**, 185303.

60. Oliveira Jr., M. H., Lopes, J. M. J., Schumann, T., Galves, L. A., Ramsteiner, M., Berlin, K., Trampert, A., and Riechert, H. (2015). Synthesis of quasi-free-standing bilayer graphene nanoribbons on SiC surfaces, *Nat. Commun.*, **6**, 7632.

61. Endo, A., Komori, F., Morita, K., Takashi, K., and Tanaka, S. (2015). Highly anisotropic parallel conduction in the stepped substrate of epitaxial graphene grown on vicinal SiC, *J. Low Temp. Phys.*, **179**, pp. 237–250.

62. Dabrowski, P., Rogala, M., Wlasny, I., Klusek, Z., Kopciuszynski, M., Jalochowski, M., Strupinski, W., and Baranowski, J. M. (2015). Nitrogen doped epitaxial graphene on 4H-SiC(0001): Experimental and theoretical study, *Carbon*, **94**, pp. 214–223.

63. Boutchich, M., Arezki, H., Alamarguy, D., Ho, K. I., Sediri, H., Günes, F., Alvarez, J., Kleider, J. P., Lai, C. S., and Ouerghi, A. (2014). Atmospheric pressure route to epitaxial nitrogen-doped trilayer graphene on 4H-SiC (0001) substrate, *Appl. Phys. Lett.*, **105**, 233111.

64. Lagoute, J., Joucken, F., Repain, V., Tison, Y., Chacon, C., Bellec, A., Girard, Y., Sporken, R., Conrad, E. H., Ducastelle, F., Palsgaard, M., Andersen, N. P., Brandbyge, M., and Rousset, S. (2015). Giant tunnel-electron injection in nitrogen-doped graphene, *Phys. Rev. B*, **91**, 125442.

65. Riedl, C., Coletti, C., Iwasaki, T., Zakharov, A. A., and Starke, U. (2009). Quasi-free-standing epitaxial graphene on SiC obtained by hydrogen intercalation, *Phys. Rev. Lett.*, **103**, 246804.

66. Walter, A. L., Jeon, K. J., Bostwick, A., Speck, F., Ostler, M., Seyller, T., Moreschini, L., Kim, Y. S., Chang, Y. J., Horn, K., and Rotenberg, E. (2011). Highly p-doped epitaxial graphene obtained by fluorine intercalation, *Appl. Phys. Lett.*, **98**, 184102.

67. Masuda, Y., Norimatsu, W., and Kusunoki, M. (2015). Formation of a nitride interface in epitaxial graphene on SiC (0001), *Phys. Rev. B*, **91**, 075421.

68. Caffrey, N. M., Armiento, R., Yakimova, R., and Abrikosov, I. A. (2015). Charge neutrality in epitaxial graphene on 6H-SiC(0001) via nitrogen intercalation, *Phys. Rev. B*, **92**, 081409(R).

69. Baringhaus, J., Stöhr, A., Forti, S., Starke, U., and Tegenkamp, C. (2015). Ballistic bipolar junctions in chemically gated graphene ribbons, *Sci. Rep.*, **5**, 9955.

70. Nevius, M. S., Conrad, M., Wang, F., Celis, A., Nair, M. N., Taleb-Ibrahimi, A., Tejeda, A., and Conrad, E. H. (2015). Semiconducting graphene from highly ordered substrate interactions, *Phys. Rev. Lett.*, **115**, 136802.

71. Tang, Q. and Zhou, Z. (2013). Graphene-analogous low-dimensional materials, *Prog. Mater. Sci.*, **58**, pp. 1244–1315.

72. Shin, H.-C., Jang, Y., Kim, T. H., Lee, J. H., Oh, D. H., Ahn, S. J., Lee, J. H., Moon, Y., Park, J. H., Yoo, S. J., Park, C. Y., Whang, D., Yang, C. W., and Ahn, J. R. (2015). Epitaxial growth of a single-crystal hybridized boron nitride and graphene layer on a wide-band gap semiconductor, *J. Am. Chem. Soc.*, **137**, pp. 6897–6905.

73. Chubarov, M., Pedersen, H., Högberg, H., Czigany, Zs., and Henry, A. (2014). Chemical vapour deposition of epitaxial rhombohedral BN thin films on SiC substrates, *Cryst. Eng. Comm.*, **16**, pp. 5430–5436.

74. Younes, G., Ferro, G., Soueidan, M., Brioude, A., Souliere, V., and Cauwet, F. (2012). Deposition of nanocrystalline translucent h-BN films by chemical vapour deposition at high temperature, *Thin Solid Films*, **520**(7), pp. 2424–2428.

75. Kobayashi, Y., Hibino, H., Nakamura, T., Akasaka, T., Makimoto, T., and Matsumoto, N. (2007). Boron nitride thin films grown on graphitized 6H-SiC substrates by metalorganic vapor phase epitaxy, *Jpn. J. Appl. Phys.*, **46**, pp. 2554–2557.

76. Hollander, M. J., Agrawal, A., Bresnehan, M. S., LaBella, M., Trumbull, K. A., Cavalero, R., Snyder, D. W., Datta, S., and Robinson, J. A. (2013). Heterogeneous integration of hexagonal boron nitride on bilayer quasi-free-standing epitaxial graphene and its impact on electrical transport properties, *Phys. Status Solidi A*, **210**(6), pp. 1062–1070.

77. Lin, Y. C., Lu, N., Perea-Lopez, N., Li, J., Lin, Z., Peng, X., Lee, C. H., Sun, C., Calderin, L., Browning, P. N., Bresnehan, M. S., Kim, M. J., Mayer, T. S., Terrones, M., and Robinson, J. A. (2014). Direct synthesis of van der Waals solids, *ACS Nano*, **8**(4), pp. 3715–3723.

78. Lin, Y. C., Chang, C. Y. S., Ghosh, R. K., Li, J., Zhu, H., Addou, R., Diaconescu, B., Ohta, T., Peng, X., Lu, N., Kim, M. J., Robinson, J. T., Wallace, R. M.,

Mayer, T. S., Datta, S., Li, L. J., and Robinson, J. A. (2014). Atomically thin heterostructures based on single-layer tungsten diselenide and graphene, *Nano Lett.*, **14**(12), pp. 6936–6941.

79. Miwa, J. A., Dendzik, M., Grønborg, S. S., Bianchi, M., Lauritsen, J. V., Hofmann, P., and Ulstrup, S. (2015). Van der Waals epitaxy of two-dimensional MoS_2-graphene heterostructures in ultrahigh vacuum, *ACS Nano*, **9**(6), pp. 6502–6510.

80. Lee II, E. W., Ma, L., Nath, D. N., Lee, C. H., Arehart, A., Wu, Y., and Rajan, S. (2014). Growth and electrical characterization of two-dimensional layered MoS_2/SiC heterojunctions, *Appl. Phys. Lett.*, **105**, 203504.

Chapter 2

Growth Mechanism, Structures, and Properties of Graphene on SiC(0001) Surfaces: Theoretical and Experimental Studies at the Atomic Scale

Wataru Norimatsu,[a] Stephan Irle,[b] and Michiko Kusunoki[c]

[a]*Department of Applied Chemistry, Nagoya University, Furo-cho, Chikusa-ku, Nagoya-shi, Aichi, 464-8603, Japan*

[b]*Institute of Transformative Bio-Molecules (WPI-ITbM) & Department of Chemistry, Nagoya University, Furo-cho, Chikusa-ku, Nagoya-shi, Aichi, 464-8602, Japan*

[c]*EcoTopia Science Institute, Nagoya University, Furo-cho, Chikusa-ku, Nagoya-shi, Aichi, 464-8603, Japan*

w_norimatsu@esi.nagoya-u.ac.jp

Graphene can be grown by thermal decomposition of SiC(0001) surfaces [1, 2]. Graphene on SiC is one of the best candidates for application in electronics because wafer-scale single crystalline graphene can be epitaxially grown on the semi-insulating substrate. For the controlled production of high-quality graphene on SiC, atomic-scale understanding of the growth process is needed. Roughly speaking, the growth process is driven by the sublimation

Epitaxial Graphene on Silicon Carbide: Modeling, Devices, and Applications
Edited by Gemma Rius and Philippe Godignon

ISBN 978-981-4774-20-8 (Hardcover), 978-1-315-18614-6 (eBook)
www.panstanford.com

of silicon atoms from the SiC surface and the subsequent rearrangement of the remaining carbon atoms into a densely packed honeycomb structure. In this chapter, we summarize the graphene growth mechanisms at the atomic scale revealed by high-resolution transmission electron microscope (HRTEM) observations combined with molecular-dynamics calculations based on density-functional tight-binding (DFTB) method. The graphene growth mechanism differs on different crystallographic surfaces, such as the Si- and C-faces. We also describe the possible atomic structures of graphene on SiC and their basic electronic properties.

2.1 Introduction

Graphene is a one atom–thick layer of graphite. As an ideal two-dimensional carbon material, graphene has interesting features from the viewpoint of fundamental physics as well as potential for many applications [3–5]. Particularly, graphene grown by thermal decomposition of SiC is expected to be of relevance for application in electronics because wafer-scale single crystalline graphene can be grown directly on the semi-insulating substrate [1, 2].

Figure 2.1a shows an HRTEM image of a monolayer graphene on 6H-SiC (0001) [6, 7]. In the HRTEM image, graphene can be identified as a dark line contrast. In addition to graphene, there is another carbon layer, represented by a broken line in Fig. 2.1a. In this layer, the in-plane atomic arrangement is almost the same as that of graphene, but some carbon atoms have covalent bonds with the silicon atoms below. This layer does not have the linear band dispersion characterizing graphene and does not freely contribute to the electric conduction. It is called a buffer layer [8].

Graphene on SiC is formed by selective sublimation of atomic silicon from the SiC surface at high temperature, and the subsequent rearrangement of the carbon atoms, as shown in Fig. 2.1b. In the graphene growth process, the buffer layer forms first. Further decomposition provides carbon atoms under the buffer layer, and they form a new buffer layer. The original buffer layer loses its bonding with SiC substrate and then turns into (freestanding) graphene. Graphene on SiC is thus epitaxially formed. The primary concept of the terminology "buffer layer" was based on its location, which is between graphene and SiC [9, 10]. In these days, however, the first carbon layer is commonly called a buffer layer even when no graphene is formed onto it [11]. The right image shown in Fig. 2.1b is bilayer graphene, formed by the iteration of this process [7, 12, 13].

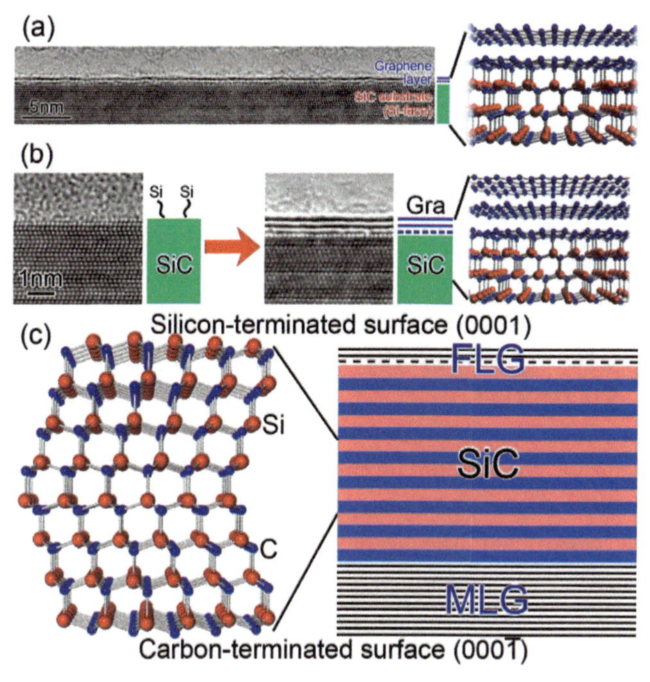

Figure 2.1 Structure of graphene on SiC(0001): (a) Monolayer graphene on SiC (0001); the blue broken line corresponds to the buffer layer. (b) Schematic diagram of graphene grown by thermal decomposition of SiC. (c) Features of graphene on the Si-terminated (0001) and C-terminated (0001) surfaces. Figures reprinted from Refs. [7] and [13], with permission from IOP Publishing, and Ref. [12], with permission from Royal Society of Chemistry.

The SiC(0001) surfaces, which are suitable for graphene growth, include the Si-terminated (0001) and the C-terminated (0001) faces. The above results were obtained for graphene on the Si-face. Figure 2.1c shows the crystal structure of 6H-SiC, its surfaces, and the main features of graphene grown on them [7, 12].

The structure of SiC is characterized by the stacked Si–C bilayer, where silicon and the carbon atoms arrange in a honeycomb-like structure. The stacking sequence of the 6H-SiC is of ABCACBA type.

In a similar fashion, we can use the 4H- and 3C-SiC with ABCBA- and ABCA-type stacking, respectively. Among these polytypes, the features of graphene grown on them are quite similar. On the other hand, graphene is significantly different when grown either on the SiC Si- or C-faces. As shown in Fig. 2.1c, monolayer or bilayer graphene (few-layer graphene, FLG) can be grown with a high controllability on the Si-face, and typically multilayer graphene (MLG) can be grown on the C-face. This difference is due to varying reactivities of the two types of surfaces, resulting in different graphene growth mechanisms. In this chapter, we focus on the growth mechanism of graphene on these two distinctive faces combining experimental and theoretical results. Afterward, we also describe the structural features and the basic electronic properties of graphene on SiC.

2.2 Growth Mechanism of Graphene on SiC(0001)

Graphene growth on SiC occurs by the thermal decomposition of the SiC surface and by atomic carbon rearrangement on the surface. Historically, the graphitization phenomenon induced by the thermal decomposition of SiC was revealed by Acheson in 1896, who produced carborundum (SiC) [14]. A crystallographic study of graphite grown from SiC by an X-ray diffraction technique was performed by Badami in 1965 [15]. The detailed experiments of van Bommel revealed the crystallographic orientation relation between graphite and SiC in 1975 [16]. In 1998, Forbeaux et al. used low-energy electron diffraction to show that a carbon reconstructed layer with a $6\sqrt{3} \times 6\sqrt{3}R30$ (6R30) structure was first formed on the surface, which is the buffer layer as stated above [17]. Further decomposition produces free carbon atoms to form the next 6R30 layer beneath the present layer, and the original 6R30 layer turns into graphene. The second 6R30 layer then becomes the new buffer layer. The direct observation of few-layer graphene on SiC (0001) was carried out in 2000 [18].

Beyond the structural investigation of graphene growth, we should also focus on the surface morphology of SiC. The surface structure and the morphology play a significant role in the graphene growth, because the carbon rearrangement on the surface is its essence. The SiC is formed by the stacking of the Si–C bilayers, and the SiC(0001) surface contains atomic steps with a bilayer height of 2.5 Å adjacent to the flat terrace region. On the step and the terrace, the stability of the surface atoms against the thermal decomposition is significantly different because the number of dangling bonds differs. It is then reasonable to expect that these differences between steps and terraces bear a significant effect on the graphene growth mechanism. In this section, we review the experimental and theoretical results of the graphene growth mechanisms on the Si- and the C-faces.

2.1.1 Experimental Results Concerning the Mechanism on the Si-Face (0001)

Experimental results from the growth on the Si-face are shown in Fig. 2.2. The HRTEM images show the early stage of graphene layer growth. The detailed experimental procedure was described in Ref. [19]. In Fig. 2.2a, about four graphene layers with a width of 1~2 nm are formed, covering the SiC steps. Here, this graphene nucleus is anchored via σ-bonds on the lower terrace. No clear contrast was observed in other areas. This suggests that graphene nucleates around the SiC steps. In the images in panels (b–d), graphene "standing" in this way on the right lower terrace spreads to the left. Around the step present on the left side, the number of graphene layers changes. This fact indicates that graphene nucleates at the right-side step and grows toward the left side, and the growth of partial layers occurs on the terrace of the left-side step. Thus, graphene growth concurs with step retreat. In the image in panel (e), a somewhat poorly resolved trilayer graphene is present around the arrowed step, and five or six layers can be found on the left and right sides, suggesting the coalescence of these layers. After the growth completes, graphene covers the step, as shown in panel (f).

This growth mechanism is illustrated schematically in Fig. 2.3. First, the Si atoms at the step edge preferentially sublime, and the graphene nucleates, covering the step. Then, this nucleus grows from the step horizontally on the upper terrace. The same sequence of events occurs on the lower terrace, eventually leading up to coalescence. Graphene, finally, covers the whole surface of the SiC substrate, including the step edges. Similar results have been reported by several other research groups [20–24]. After one layer covers the surface, it represents a barrier for further decomposition. Consequently, the growth rate of the second layer substantially decreases [25]. In order to form uniform monolayer graphene, the first layer should complete its growth before the second layer starts to grow. This can be controlled by the step-density and the decomposition speed.

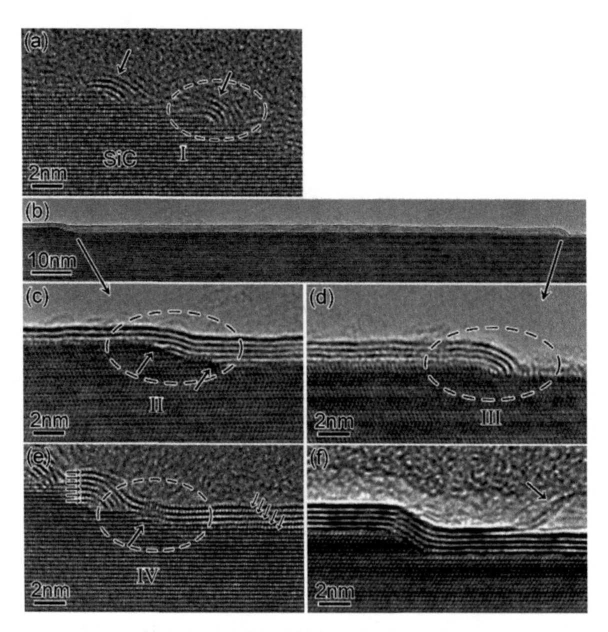

Figure 2.2 HRTEM images corresponding to (a) graphene nucleation and (b–f) growth stages on SiC (0001). Reprinted with permission from Ref. [19], Copyright 2010, Elsevier.

Graphene growth from SiC is different from typical step-flow crystal growth in terms of the fact that it grows on the upper terrace after nucleation at the step [13]. This is demonstrated in Fig. 2.4. In the image in panel (a), there is a dimple in the center, where no graphene is found, and graphene layers are formed on the upper layers on the left and the right side of the dimple. This observation suggests that graphene, nucleated at the step edges, grows toward the upper terraces, as illustrated in panels (b) and (c). As mentioned, this is different from the normal step-flow growth where the crystal grows onto the lower terrace from the step. Similar behaviors were revealed also by atomic force microscope (AFM) observations. Figures 2.4d and 2.4e are the topography and the phase image of the same area. The AFM phase image owes its contrast to the viscoelasticity of the surface material(s), where a difference in the number of graphene layers is revealed as changes in phase signal contrast. In this example, it was found that the bright and the dark areas correspond to trilayer and bilayer graphene [12]. In these AFM images, most of the bright areas can be found on the upper terrace for step edges marked by red broken lines. This clearly indicates that graphene grows on the upper terrace away from the step edge after nucleating there, as indicated by the blue arrows, pointing toward the growth direction. In other words, graphene on the Si-face grows in a layer-by-layer manner. Thus, on the Si-face, the graphene thickness can be controlled easily, and wafer-scale homogeneous monolayer graphene can be grown if this mechanism is taken into account.

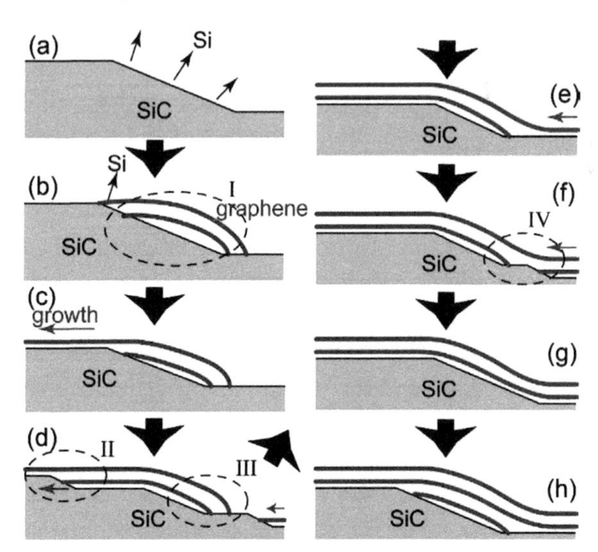

Figure 2.3 Schematic diagram of the growth mechanism of graphene on SiC (0001). Reprinted with permission from Ref. [19], Copyright 2010, Elsevier.

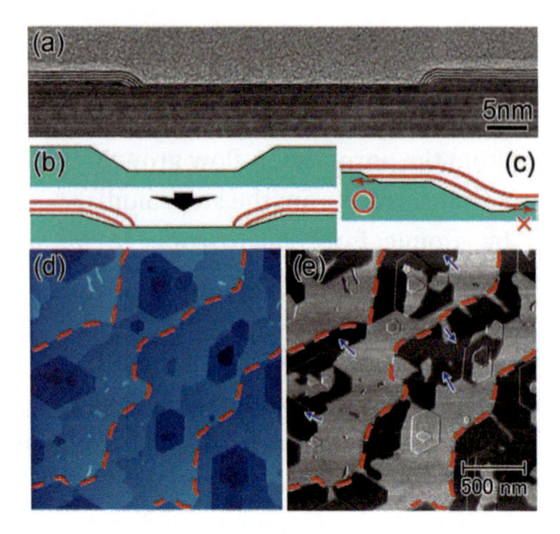

Figure 2.4 Graphene growth direction: (a–c) Graphene grows only on the upper terrace; (d, e) AFM images showing that the bright graphene region expanded in the direction indicated by blue arrows. Figures reprinted from Refs. [7] and [13], with permission from IOP Publishing.

2.2.2 Theoretical Simulations of the Growth Mechanism on the Si-Face (0001)

Along with the experimental reports, theoretical studies on the graphene growth mechanism on the Si-face have been performed. Graphene growth from SiC is the consequence of silicon sublimation and carbon rearrangement. Accordingly, graphene formation can be simulated by the investigation of the behavior of the carbon atoms after the removal of the silicon atoms from the SiC surface at finite temperatures, using calculations based on a reasonably accurate, yet computationally highly efficient quantum-mechanical/molecular-dynamics framework [21]. Here, the (initial) surface structure of SiC was carefully taken into account, including both the decomposition of the step-less terrace and the step-preferential decomposition.

The calculation results for the terrace are summarized in Fig. 2.5. Details of the calculation method are described in Ref. [26]. Using the initial structure shown in Fig. 2.5a, the dynamics of the carbon atoms after stepwise removal of silicon atoms at 1500 K was simulated. Silicon atoms were removed in the order indicated by circled numbers and arrows, with an interval of 1 ps for each Si atom removal to allow crystal relaxation. Figure 2.5b displays the atomic arrangement after one bilayer decomposition. The carbon atoms formed a chain, and the number of chains increased in panel (c). In panel (d), there are a few five- or six-membered rings, but most of the carbon atoms take a chain form, and some of the chains "stand up" on the surface. It should be noted here that the number of carbon atoms required for the deposition of a graphene monolayer can be provided by at least the decomposition of three Si–C bilayers. Then, after the three bilayer decomposition as shown in panel (d), the number of carbon atoms is sufficient for monolayer graphene formation. However, Fig. 2.5d shows no sign of graphene-like structure, which suggests that some other factor is necessary for graphene formation, such as a sufficiently long annealing time.

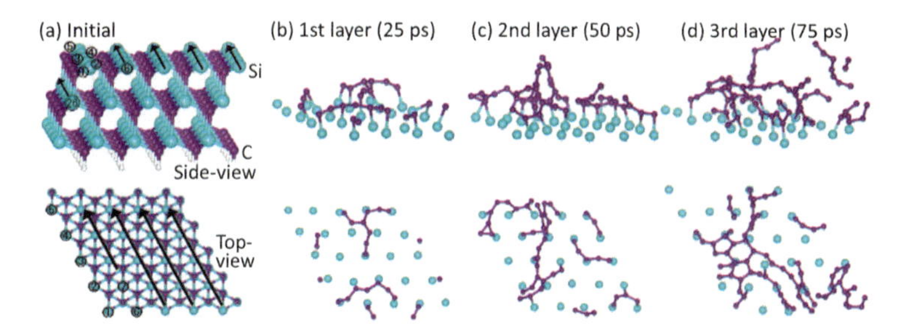

Figure 2.5 DFTB simulation results of graphene nucleation in the terrace model: (a) initial structural model and (b–d) snapshots after 1, 2, and 3 bilayer decomposition. Reprinted with permission from Ref. [26], Copyright 2013, AIP Publishing LLC.

Subsequently, the effect of step edges was also investigated. Figure 2.6a shows the model initial structure. After 30 ps (b), many carbon chains were formed like in the case of the terrace model in Fig. 2.5b,c, and after 60 ps in panel (c), some ring structures can be seen. After 90 ps, as shown in panel (d), there are regions with many fused carbon rings. A statistical analysis showed that the number of five-, six-, or seven-membered rings of the step model was higher than that of the terrace model [26]. This fact indicates that the step edge atoms acted as the carbon-trapping sites and encouraged the fused carbon ring formation. These events are consistent with experimental observations, as mentioned earlier.

The growth dynamics after nucleation was also theoretically investigated. From the initial structure shown in Fig. 2.7a, the silicon atoms were removed one by one, and then the situation of the growth was simulated. As shown in Fig. 2.7b–d, free carbon atoms produced after the silicon removal connected to the edge of the carbon network, leading to graphene growth. In some situations, the residual carbon created a chain, which would be a nucleation center for the next graphene layer. The reaction dynamics shown in Figs. 2.5–2.7 can be viewed as movies that can be downloaded from the journal website of Ref. [26]. Similar theoretical results emphasizing the importance of the surface step during graphene growth on SiC were also reported by several other groups [27, 28].

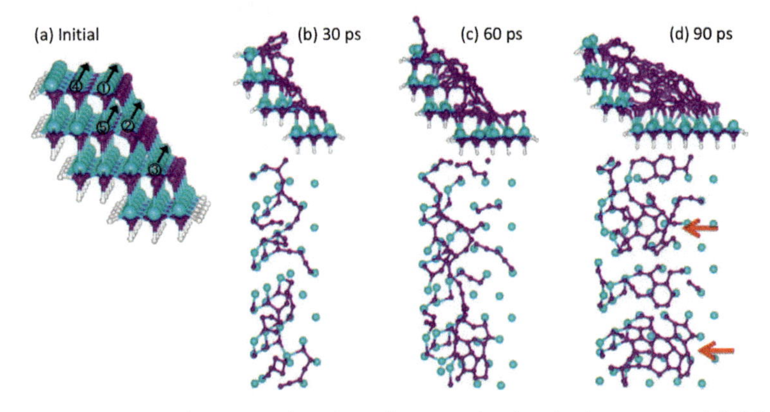

Figure 2.6 Simulation results of graphene nucleation in the step model: (a) initial structure and (b–d) snapshots after 30, 60, and 90 ps. Reprinted with permission from Ref. [26], Copyright 2013, AIP Publishing LLC.

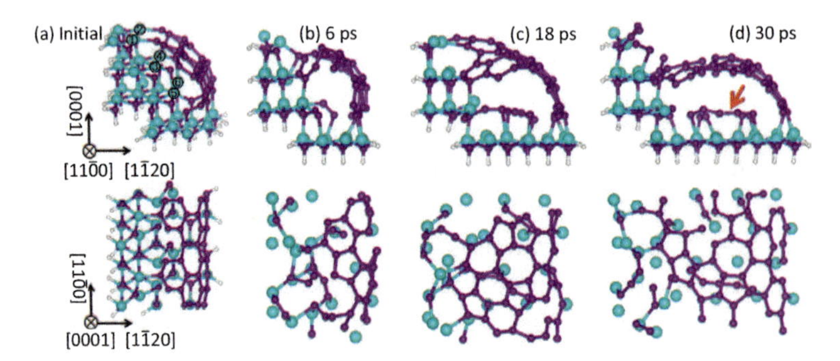

Figure 2.7 Simulation results of graphene growth after nucleation: (a) initial structure and (b–d) snapshots after 6, 18, and 30 ps. Reprinted with permission from Ref. [26], Copyright 2013, AIP Publishing LLC.

2.2.3 Experimental Results Concerning the Mechanism on the C-Face (0001)

In general, the Si- and C-terminated surfaces have different properties. In particular, the difference in the surface reactivity plays an important role for the graphene growth mechanism. Specifically, it is known that the reactivity of the C-face is higher than that of the Si-face [29, 30]. Figure 2.8 shows the HRTEM images corresponding to the graphene growth process on the C-face [31]. On the C-face, the initial graphene nucleation happens inside the void with the diameter of 3~5 nm and the depth of about 1 nm, which was probably created by a local decomposition of the surface. In the crater, several-layered graphene undergoes nucleation. After the nucleation, the area of the crater increases, which accompanies graphene growth, as shown in panels (c) and (d). Simultaneously, the area of the knoll region, where no graphene could be found, decreases and it finally vanishes, as shown in panel (f). Further decomposition leads to the increase in the number of layers. In other words, the number of layers at the nucleation stage rarely changes until graphene covers the surface completely.

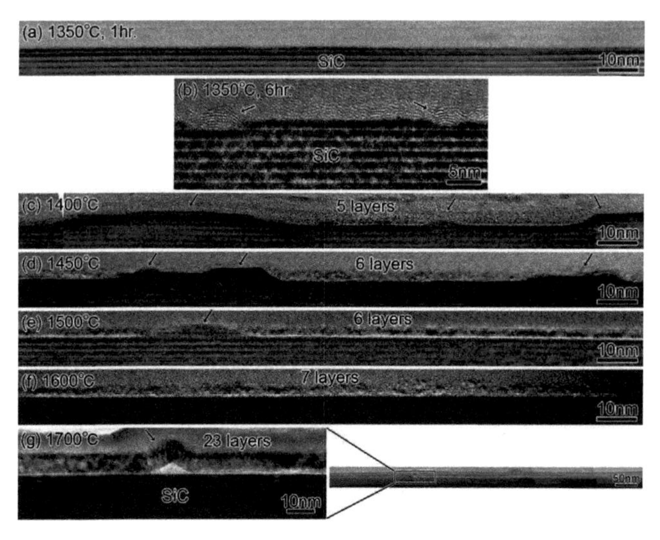

Figure 2.8 HRTEM images of graphene nucleation and growth stages on SiC (0001). Reprinted with permission from Ref. [31], Copyright 2011, American Physical Society.

A schematic diagram of the growth mechanism on the C-face is shown in Fig. 2.9. On the C-face, decomposition starts not only at the steps but also at any sites on the terrace, forming holes with several layers of graphene on it. The void then spreads in all directions of the surface, acting as a driver for the graphene growth. After graphene covers the surface, the decomposition direction changes to be normal to the surface and, therefore, the number of graphene layers increases. Importantly, the larger the number of layers, the higher the probability to form ripples of graphene.

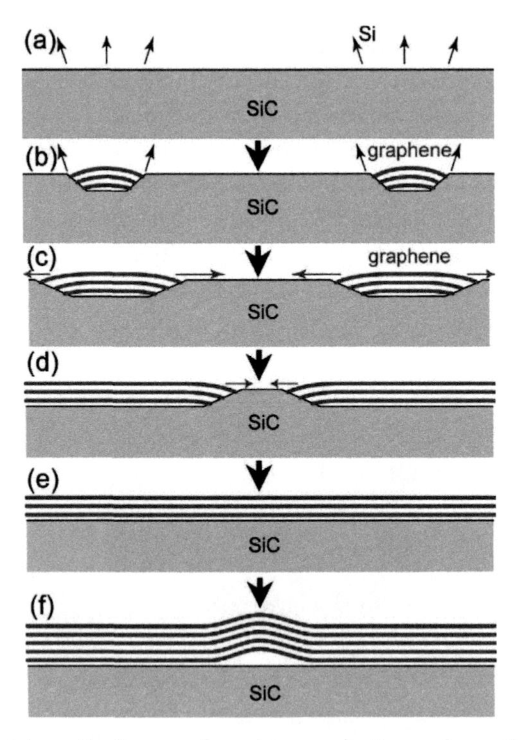

Figure 2.9 Schematic diagram of graphene nucleation and growth mechanism on SiC (0001). Reprinted with permission from Ref. [31], Copyright 2011, American Physical Society.

Similar results were obtained by several other research groups [32–34], which confirms that the more reactive C-face decomposes not only at the steps but also on the terraces, while the Si-face with lower reactivity decomposes only/mainly at the steps [31]. In addition, the C-face graphene has a tendency to grow in a multilayer form. This is because the number of layers at the nucleation stage is steady until full coverage and does not follow the layer-by-layer growth mechanism seen on the Si-face. Thus, in order to control the graphene thickness, the number and distribution of layers at the nucleation stage should be precisely uniform.

2.2.4 Theoretical Simulations of the Growth Mechanism on the C-Face (0001)

The theoretical investigation of the decomposition process of the C-face has also been reported [35]. The initial structures are equivalent to Figs. 2.5a and 2.6a, but they were converted to the C-terminated surface. Figure 2.10 shows the decomposition of the terrace model. After the first layer decomposition, the carbon atoms form a chain-like structure. During the second layer decomposition, the chains started to form some fused rings, and finally many five-, six-, and seven-membered rings constructed an sp^2-carbon network after the third layer decomposition. This situation is far different from that on the Si-face. On the Si-face terrace, carbon chains were formed, but they did not rearrange a network. That is why, for graphene growth on the Si-face, the presence of the surface steps was quite important. These results suggest that graphene growth on the C-face does not always require the presence of surface steps.

Figure 2.11 shows the structure after 90 silicon atoms were removed from the step model. Several rings connect with each other in this case. Around the step, graphene can be nucleated, which is similar to the phenomenon on the Si-face. These calculation results are consistent with the experimental results.

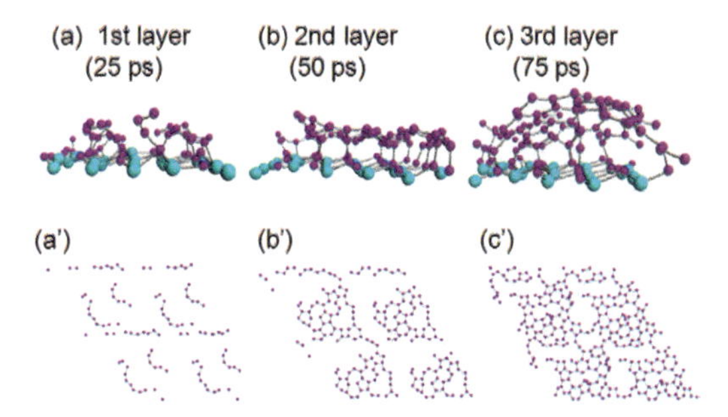

Figure 2.10 DFTB simulation results of graphene formation on the terrace of SiC (0001). The initial structure is the same as in Fig. 2.5 except for the inversion of Si and C atoms. Reprinted with permission from Ref. [35], Copyright 2014, Elsevier.

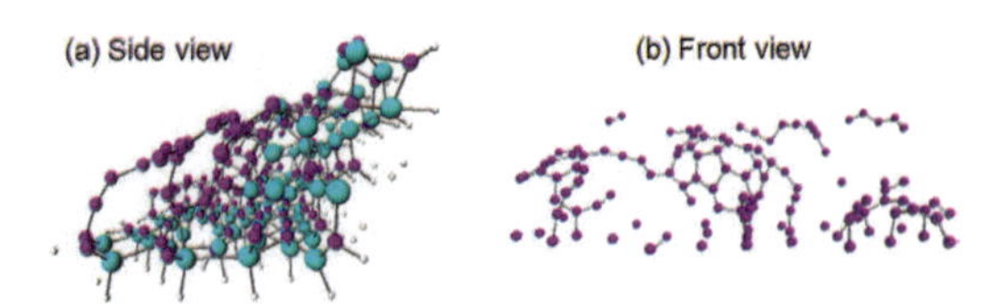

Figure 2.11 Simulation results of graphene formation on the step of SiC (0001). Reprinted with permission from Ref. [35], Copyright 2014, Elsevier.

The essential difference between the two faces is in the environment around the carbon atoms remaining after the silicon removal. On the Si-face, three carbon atoms with dangling bonds in close proximity were left, and when they bond with each other, a horizontal carbon connection is formed. On the other hand on the C-face, three carbon atoms and one other carbon atom in the second bilayer also lose their bond with silicon. The latter bond is perpendicular to the surface. This difference is one of the origins of the different growth mechanisms on the Si- and the C-faces [36]. It is, therefore, apparent that the Si-face is more suitable for graphene growth as it supports the growth of horizontal C–C bonds [36]. As a matter of fact, and not surprisingly, a homogeneous monolayer graphene can be grown on the Si-face much more easily than the C-face.

2.3 Structures and Properties of Graphene on SiC

Based on the aforementioned graphene growth mechanism, the optimization of the growth condition is possible. As graphene on the Si-face grows in a layer-by-layer manner, the Si-face substrate is usually used for homogeneous monolayer graphene growth. In this section, the structural features and the basic electronic properties, mainly of graphene on the Si-face, are discussed.

On the Si-face, monolayer graphene can be formed at relatively lower temperature, and the thickness increases with increasing growth temperature. Figure 2.12 shows the typical HRTEM images of our graphene layer(s). As shown in the images, monolayer, bilayer, and trilayer graphene can be observed as a straight and continuous line contrast, while there are defects and discontinuities in the eight-layer graphene. In fact, from our experience, almost defect-free graphene can be grown up to about six layers. When the number of layers increases, the stacking sequence of layers should be critical for the electronic structure and the properties. Among the graphite stacking structures, the AA-, AB- (Bernal), ABC- (rhombohedral), and turbostratic (TS) stackings are known. In bulk graphite, AB-, ABC-, and TS-stacking coexist with a ratio of 80:14:6 [37, 38]. Figure 2.13 is the HRTEM images showing the stacking structures of multilayer graphene on SiC. In the images in panels (a) and (b), the arrangement of the bright dots (yellow circles) is the direct evidence for the ABC-type stacking, indicating that graphene on the Si-face selectively exhibits the rhombohedral ABC-stacking [39]. Instead, on the C-face, the bright dots do not form a regular array, indicating the mixture of the stacking structures or the rotational stacking, as shown in panel (c) [13]. In some areas of graphene on the C-face, an amorphous layer of about 1 nm thickness can be observed, as shown in panel (d). Thus, the ABC-stacked graphene and the turbostratic graphene are formed on the Si- and the C-face, respectively. Multilayer graphene with rotational stacking on the C-face was reported by several groups, and the rotation gives rise to interesting modifications of the electronic properties of graphene [40–43]. The rhombohedral ABC-stacking graphene has a distinct electronic structure, which induces an electric field–induced bandgap opening, a peculiar quantum Hall effect, and a ferrimagnetic spin arrangement [44–48].

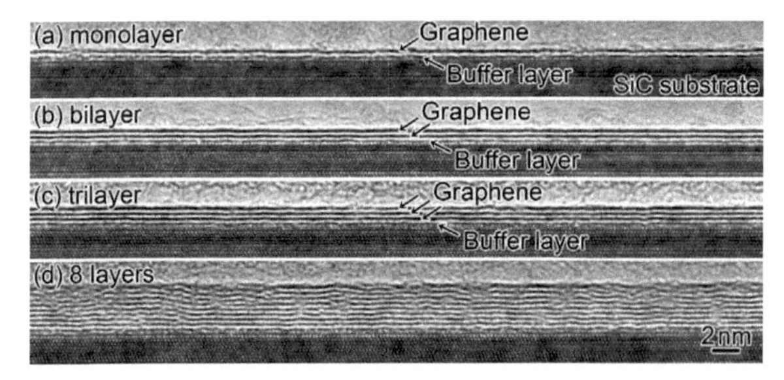

Figure 2.12 HRTEM images of graphene layer(s) on the Si-terminated SiC (0001) face. On the Si-face, graphene is present always on the buffer layer. Reprinted with permission from Ref. [6], Copyright 2008, Elsevier.

In the early days, graphene growth from SiC was performed in vacuum, such as in an ultrahigh vacuum (UHV). However, it

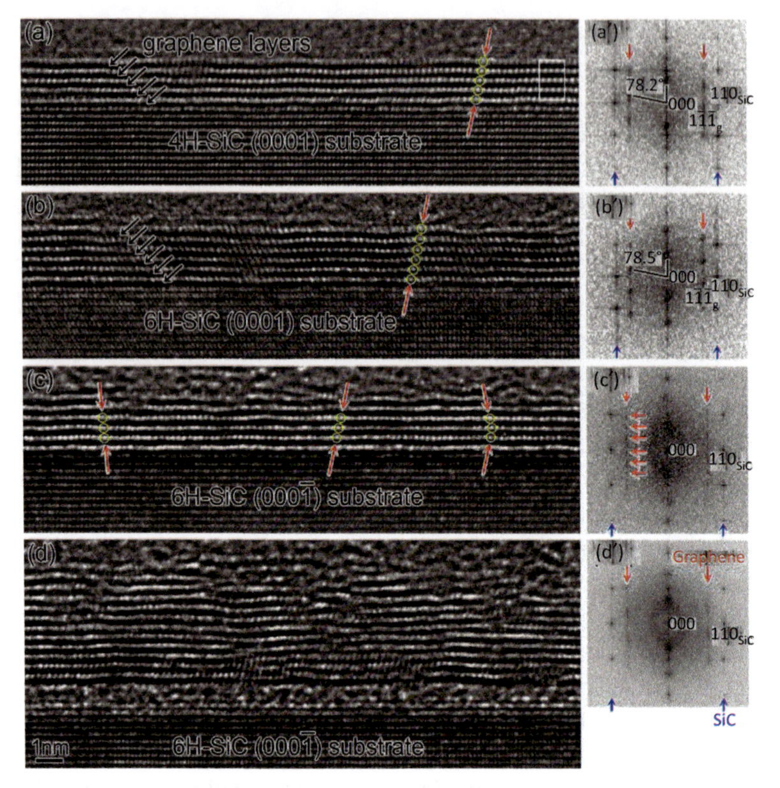

Figure 2.13 Stacking structures of multilayer graphene on SiC: (a, b) graphene with ABC-stacking on the Si-face; (c, d) graphene with rotational stacking and with the amorphous interface on the C-face. The corresponding diffraction patterns also indicate their stacking structures. Reprinted with permission from Ref. [31], Copyright 2011, and Ref. [39], Copyright 2010, American Physical Society.

was known that the UHV-grown graphene was not uniform and its surface was not flat, containing numerous pits. Around 2008, homogeneous graphene on the Si-face was obtained by growth in an Ar atmosphere [49, 50]. Prior to the graphene growth, the surface of the SiC substrate is usually cleaned by a hydrogen etching process. This process is performed to clean and flatten the surface by heating the SiC substrate in an atmosphere containing hydrogen gas. Figure 2.14a shows the AFM image of the SiC surface after hydrogen etching at 1375°C in an Ar/4%H_2 atmosphere with a gas flow rate of 1 L/min. An atomically flat surface with a step height of 1.5 nm and terrace width of about 350 nm and a highly periodic step array is obtained. Homogeneous monolayer graphene grows easily on this flat SiC surface. Figures 2.14b and 2.14c are the AFM topography and phase images of graphene grown at 1700°C, at atmospheric pressure, under Ar flow conditions. The step height is still about 1.5 nm, and the phase image has a uniform contrast, indicating a uniform number of graphene layers. The Raman spectra of this graphene sample is shown in Fig. 2.15. In the raw spectrum in panel (a), there are many peaks in the region 1000–1900 cm^{-1}, most of which come from the SiC substrate. In the spectrum after subtraction of the SiC component shown in panel (b), the sharp G and 2D bands are present at 1593 and 2710 cm^{-1}. The 2D peak consists of single component, and its full width at half maximum (FWHM) value is very small at about 36 cm^{-1}, attributable to monolayer graphene. The combined information from the HRTEM observation and the Raman spectroscopy showed that the FWHM of the 2D band less than 40 cm^{-1} was an indication of monolayer graphene. The broad peaks found in the low-wavenumber wing of the G peak and at about 1400 cm^{-1} are known to be due to the 6R30 buffer layer [52].

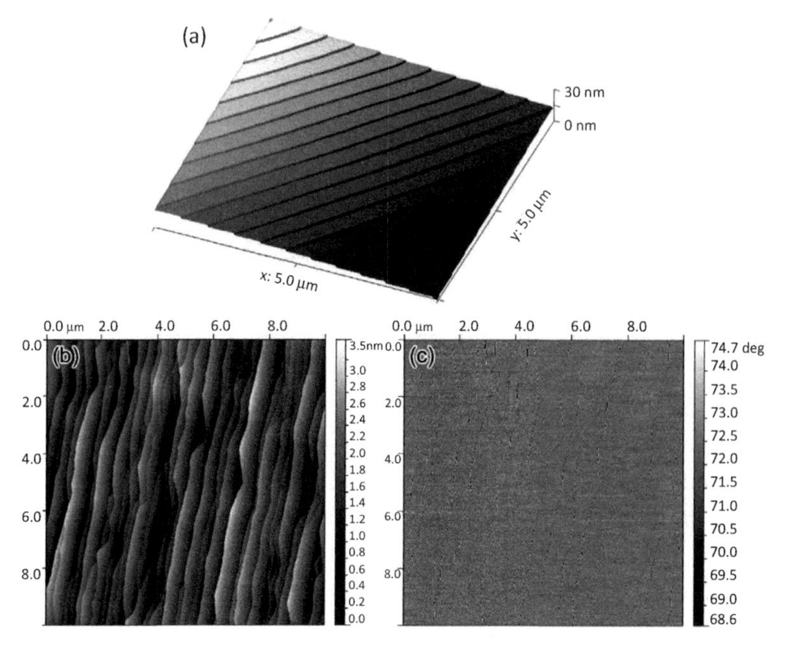

Figure 2.14 (a) AFM image of hydrogen-etched SiC surface (Si-face). (b, c) AFM topography and phase images of monolayer graphene on SiC (0001) [51].

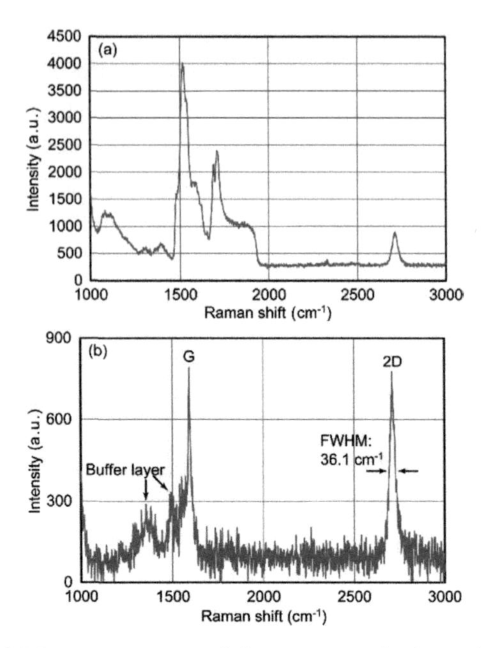

Figure 2.15 (a) Raman spectrum of the same sample shown in Fig. 2.14. (b) Spectrum after subtraction of the SiC substrate component [51].

As for the basic electronic properties, Hall-effect measurements were carried out by using the van der Pauw method in a 5 mm × 5 mm homogeneous monolayer graphene sample at temperatures between 20 K and 300 K. The results are shown in Fig. 2.16. The sheet resistance decreases with decreasing temperature. For the entire temperature range, the Hall coefficient has a negative value, indicating electron conduction. The electron concentration is about 1×10^{13} cm^{-2}, with almost no temperature dependence. This carrier concentration is relatively high, compared with the typical value for carbon materials. The electron doping is attributed to the spontaneous polarization of the hexagonal SiC substrate and the presence of the buffer layer [53]. The electron mobility is about 1800 cm^2/Vs at 20 K, which decreases with increasing temperature, resulting in about 900 cm^2/Vs at room temperature. Generally, it is known that the charge carrier mobility of graphene strongly depends on the carrier concentration. In particular, on a polar surface such as SiC, the mobility is proportional to the inverse of the square root of the carrier concentration [54, 55]. This indicates that, if we reduce the carrier concentration, the mobility value could be easily improved [56–58]. For example, a mobility of 1800 $cm^2/V\,s$ at 1×10^{13} cm^{-2} corresponds to a value of 18,000 $cm^2/V\,s$ at 1×10^{11} cm^{-2}, which is considered high mobility. This high mobility value is also a measure of the high crystal quality of epitaxial graphene on SiC, to be thus used for quantum transport experiments [5]. The mobility decrease with decreasing temperature is mainly due to the phonon scattering at the interface, due to the buffer layer [55]. This suggests that interface modification would further improve the mobility properties. In fact, interface modification by atomic intercalation at the interface successfully reduced the interface carrier scattering and modified the electronic structure of graphene [11, 59–61].

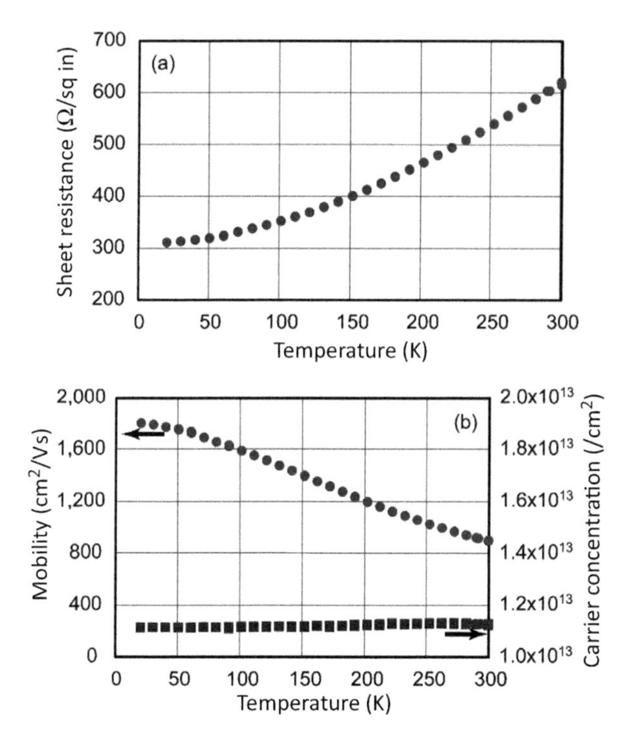

Figure 2.16 Hall-effect measurement results of monolayer graphene on SiC (0001). Temperature dependences of the (a) sheet resistance, (b) mobility and electron concentration are shown [51].

2.4 Conclusion

In this chapter, we summarized the atomic-scale growth mechanisms of graphene grown by thermal decomposition of SiC. The growth mechanism strongly depends on the surface structure and morphology. On the Si-terminated SiC surface, graphene nucleates at the atomic steps and then grows laterally on the upper terrace. On the C-face, graphene nucleates not only at the step, but is also formed on the terrace, and grows in all directions across the surface. Thanks to the layer-by-layer growth mechanism on the Si-face, wafer-scale single crystalline homogeneous monolayer graphene can be grown. The basic electronic properties of a 5×5 mm^2 sample of monolayer graphene, together with its scattering mechanism, were reported and discussed.

Acknowledgments

This work was partially supported by MEXT/JSPS KAKENHI Grant Numbers 25107002 and 26706014, and the Asahi Glass Foundation. S.I. acknowledges financial support from a CREST grant by JST.

References

1. C. Berger, Z. Song, T. Li, X. Li, Y. Asmerom, Y. Ogbazghi, R. Feng, Z. Dai, A. N. Marchenkov, E. H. Conrad, P. N. First, and W. de Heer, *J. Phys. Chem.*, **108**, 19912 (2004).
2. C. Berger, Z. Song, X. Li, X. Wu, N. Brown, C. Naud, D. Mayou, T. Li, J. Hass, A. N. Marchenkov, E. H. Conrad, P. N. First, and W. de Heer, *Science*, **312**, 1191 (2006).
3. K. S. Novoselov, A. K. Geim, S. V. Morozov, D. Jiang, Y. Zhang, S. V. Dubonos, I. V. Grigorieva, and A. A. Firsov, *Science*, **306**, 666 (2004).
4. C. R. Dean, A. F. Young, I. Meric, C. Lee, L. Wang, S. Sorgenfrei, K. Watanabe, T. Taniguchi, P. Kim, K. L. Shepard, and J. Hone, *Nat. Nanotechnol.*, **5**, 722 (2010).
5. K. S. Novoselov, V. I. Fal'ko, L. Colombo, P. R. Gellert, M. G. Schwab, and K. Kim, *Nature*, **490**, 192 (2012).
6. W. Norimatsu and M. Kusunoki, *Chem. Phys. Lett.*, **468**, 52 (2009).
7. W. Norimatsu and M. Kusunoki, *Semicond. Sci. Technol.*, **29**, 064009 (2014).

8. S. Kim, J. Ihm, H. J. Choi, and Y.-W. Son, *Phys. Rev. Lett.*, **100**, 176802 (2008).
9. A. Mattausch and O. Pankratov, *Phys. Rev. Lett.*, **99**, 076802 (2007).
10. F. Varchon, R. Feng, J. Hass, X. Li, B. Ngoc Nguyen, C. Naud, P. Mallet, J.-Y. Veuillen, C. Berger, E. H. Conrad, and L. Magaud, *Phys. Rev. Lett.*, **99**, 126805 (2007).
11. C. Riedl, C. Colleti, T. Iwasaki, A. A. Zakharov, and U. Starke, *Phys. Rev. Lett.*, **103**, 246804 (2009).
12. W. Norimatsu and M. Kusunoki, *Phys. Chem. Chem. Phys.*, **16**, 3501 (2014).
13. W. Norimatsu and M. Kusunoki, *J. Phys. D Appl. Phys.*, **47**, 094017 (2014).
14. E. G. Acheson, Patent US568323 (1896).
15. D. V. Badami, *Carbon*, **3**, 53 (1965).
16. A. J. van Bommel, J. E. Crombeen, A. van Tooren, *Surf. Sci.*, **48**, 463 (1975).
17. I. Forbeaux, J.-M. Themlin, and J.-M. Debever, *Phys. Rev. B*, **58**, 16396 (1998).
18. M. Kusunoki, T. Suzuki, T. Hirayama, N. Shibata, and K. Kaneko, *Appl. Phys. Lett.*, **77**, 531 (2000).
19. W. Norimatsu and M. Kusunoki, *Physica E*, **42**, 691 (2010).
20. J. Robinson, X. Weng, K. Trumbull, R. Cavalero, M. Wetherington, E. Frantz, M. LaBella, Z. Hughes, M. Fanton, and D. Snyder, *ACS Nano*, **4**, 153 (2010).
21. T. Ohta, N. C. Bartelt, S. Nie, K. Thurmer, and G. L. Kellogg, *Phys. Rev. B*, **81**, 121411(R) (2010).
22. M. Hupalo, E. H. Conrad, and M. C. Tringides, *Phys. Rev. B*, **80**, 041401(R) (2009).
23. V. Borovikov and A. Zangwill, *Phys. Rev. B*, **80**, 121406(R) (2009).
24. M. L. Bolen, S. E. Harrison, L. B. Biedermann, and M. A. Capano, *Phys. Rev. B*, **80**, 115433 (2009).
25. S. Tanaka, K. Morita, and H. Hibino, *Phys. Rev. B*, **81**, 041406(R) (2010).
26. M. Morita, W. Norimatsu, H.-J. Qian, S. Irle, and M. Kusunoki, *Appl. Rev. Lett.*, **103**, 141602 (2013).
27. F. Ming and A. Zangwill, *Phys. Rev. B*, **84**, 115459 (2011).
28. M. Inoue, Y. Kangawa, K. Wakabayashi, H. Kageshima, and K. Kakimoto, *Jpn. J. Appl. Phys.*, **50**, 038003 (2011).

29. H. Matsunami and T. Kimoto, *Mater. Sci. Eng. R*, **20**, 125 (1997).
30. E. A. Ray, J. Rozen, S. Dhar, L. C. Feldman, and J. R. Williams, *J. Appl. Phys.*, **103**, 023522 (2008).
31. W. Norimatsu, J. Takada, and M. Kusunoki, *Phys. Rev. B*, **84**, 035424 (2011).
32. J. K. Hite, M. E. Twigg, J. L. Tedesco, A. L. Friedman, R. L. Myers-Ward, C. R. Eddy Jr., and D. K. Gaskill, *Nano Lett.*, **11**, 1190 (2011).
33. F. J. Ferrer, E. Moreau, D. Vignaud, D. Deresmes, S. Godey, and X. Wallart, *J. Appl. Phys.*, **109**, 054307 (2011).
34. R. Zhang, Y. Dong, W. Kong, W. Han, P. Tan, Z. Liao, X. Wu, and D. Yu, *J. Appl. Phys.*, **112**, 104307 (2012).
35. N. Ogasawara, W. Norimatsu, S. Irle, and M. Kusunoki, *Chem. Phys. Lett.*, **595–596**, 266 (2014).
36. Z. Wang, S. Irle, G. Zheng, M. Kusunoki, and K. Morokuma, *J. Phys. Chem.*, **111**, 12960 (2007).
37. R. R. Haering, *Can. J. Phys.*, **36**, 352 (1958).
38. J.-C. Charlier, X. Gonze, and J.-P. Michenaud, *Carbon*, **32**, 289 (1994).
39. W. Norimatsu and M. Kusunoki, *Phys. Rev. B*, **81**, 161410(R) (2010).
40. J. Kuroki, W. Norimatsu, and M. Kusunoki, *e-J. Surf. Sci. Nanotech.*, **10**, 396 (2012).
41. J. Hass, F. Varchon, J. E. Millan-Otoya, M. Sprinkle, N. Sharma, W. A. de Heer, C. Berger, P. N. First, L. Magaud, and E. H. Conrad, *Phys. Rev. Lett.*, **100**, 125504 (2008).
42. L. I. Johansson, S. Watcharinyanon, A. A. Zakharov, T. Iakimov, R. Yakimova, and C. Virojanadara, *Phys. Rev. B*, **84**, 125405 (2011).
43. I. Brihuega, P. Mallet, H. Gonzalez-Herrero, G. Trambly de Laissardiere, M. M. Ugeda, L. Magaud, J. M. Gomez-Rodrıguez, F. Yndurain, and J.-Y. Veuillen, *Phys. Rev. Lett.*, **109**, 196802 (2012).
44. S. Latil and L. Henrard, *Phys. Rev. Lett.*, **97**, 036803 (2006).
45. M. Aoki and H. Amawashi, *Solid State Commun.*, **142**, 123 (2007).
46. M. F. Craciun, S. Russo, M. Yamamoto, J. B. Oostinga, A. F. Morpurgo, and S. Tarucha, *Nat. Nanotechnol.*, **4**, 383 (2009).
47. M. Otani, M. Koshino, Y. Takagi, and S. Okada, *Phys. Rev. B*, **81**, 161403(R) (2010).
48. C. Coletti, S. Forti, A. Principi, K. V. Emtsev, A. A. Zakharov, K. M. Daniels, B. K. Daas, M. V. S. Chandrashekhar, T. Ouisse, D. Chaussende, A. H. MacDonald, M. Polini, and U. Starke, *Phys. Rev. B*, **88**, 155439 (2013).

49. C. Virojanadara, M. Syvajarvi, R. Yakimova, L. I. Johansson, A. A. Zakharov, and T. Balasubramanian, *Phys. Rev. B*, **78**, 245403 (2008).
50. K. V. Emtsev, A. Bostwick, K. Horn, J. Jobst, G. L. Kellogg, L. Ley, J. L. McChesney, T. Ohta, S. A. Reshanov, J. Rohrl, E. Rotenberg, A. K. Schmid, D. Waldmann, H. B. Weber, and Th. Seyller, *Nat. Mater.*, **8**, 203 (2009).
51. M. Kusunoki, W. Norimatsu, J. Bao, K. Morita, and U. Starke, *J. Phys. Soc. Jpn.*, **84**, 121014 (2015).
52. F. Fromm, M. H. Oliveira Jr., A. Molina-Sanchez, M. Hundhausen, J. M. J. Lopes, H. Riechert, L. Wirtz, and Th. Seyller, *New J. Phys.*, **15**, 043031 (2013).
53. J. Ristein, S. Mammadov, and Th. Seyller, *Phys. Rev. Lett.*, **108**, 246104 (2012).
54. V. Perebeinos and Ph. Avouris, *Phys. Rev. B*, **81**, 195442 (2010).
55. A. J. M. Giesbers, P. Prochazka, and C. F. J. Flipse, *Phys. Rev. B*, **87**, 195405 (2013).
56. J. Jobst, D. Waldmann, F. Speck, R. Hirner, D. K. Maude, Th. Seyller, and H. B. Weber, *Phys. Rev. B*, **81**, 195434 (2010).
57. D. B. Farmer, V. Perebeinos, Y.-M. Lin, C. Dimitrakopoulos, and Ph. Avouris, *Phys. Rev. B*, **84**, 205417 (2011).
58. S. Tanabe, Y. Sekine, H. Kageshima, M. Nagase, and H. Hibino, *Phys. Rev. B*, **84**, 115458 (2011).
59. F. Speck, J. Jobst, F. Fromm, M. Ostler, D. Waldmann, M. Hundhausen, H. B. Weber, and Th. Seyller, *Appl. Phys. Lett.*, **99**, 122106 (2011).
60. K. V. Emtsev, A. A. Zakharov, C. Coletti, S. Forti, and U. Starke, *Phys. Rev. B*, **84**, 125423 (2011).
61. Y. Masuda, W. Norimatsu, and M. Kusunoki, *Phys. Rev. B*, **91**, 075421 (2015).

Chapter 3

Fabrication of Graphene by Thermal Decomposition of SiC

Gholam Reza Yazdi, Tihomir Iakimova, and Rositza Yakimova
Department of Physics, Chemistry and Biology, Linköping University, SE-58183 Linköping, Sweden
yazdi@ifm.liu.se

3.1 Introduction

Historically, the word graphene is derived from the Greek word *graphein*, which means "to write"—one of the earliest uses of graphene (in its bulk or macroscopic version, graphite). Single layers of carbon atoms that have been isolated from graphite are commonly referred to as "graphene," a precise definition of this material that has been available since 1986. For instance, Boehm et al. in 1962 produced thin lamellar carbon after the chemical reduction of GO. In this process, dilute alkaline media with hydrazine, hydrogen sulfide, or iron salts are used as reducing agents [1]. In 1975, van Bommel et al. sublimed silicon atoms from silicon carbide (SiC) (0001) crystal under high vacuum ($<10^{-10}$ Torr) and at elevated temperature, carbon atoms that were left behind after

Epitaxial Graphene on Silicon Carbide: Modeling, Devices, and Applications
Edited by Gemma Rius and Philippe Godignon

ISBN 978-981-4774-20-8 (Hardcover), 978-1-315-18614-6 (eBook)
www.panstanford.com

sublimation of silicon arranged themselves into a thin graphitic sheet (graphene) [2]. Monolayer flakes of graphene were intended through micromechanical approach in 1999, but this process initially failed [3]. Geim and Novoselov in 2004 used the same micromechanical approach and finally obtained thin flakes of carbon by pressing highly oriented pyrolytic graphite (HOPG) surface against silicon wafer [4]. Geim and Novoselov were awarded the Noble Prize in Physics in 2010 for their demonstration of the unique and extraordinary electronic properties of graphene; indeed, 2010 is called "the year of graphene."

Since graphene discovery, the unconventional two-dimensional electron gas properties of graphene has attracted remarkable interest for condensed physics and materials science, as well as the advance of a number of technologies. Graphene is a two-dimensional crystal of honeycomb structure of sp^2-bonded carbon atoms; the carbon–carbon distance is 1.42 Å and the lattice constant is 2.46 Å (Fig. 3.1a,b), which can exist as a freestanding state. Nowadays, the concept "graphene" is not strictly used for single layers, but also for bilayer and few-layer graphene (3 to <10 layers), which all can be viewed as three different types of two-dimensional crystals [5]. Graphene has extremely high electric and thermal conductivity, optical transparency, single-molecule gas detection sensitivity, high electron mobility, and mechanical toughness [6]. Optoelectronically, it is a semi-metal with linear band dispersion around the Dirac point (Fig. 3.1c) due to its two-dimensional honeycomb structure [6, 7]. This makes the material to behave differently from conventional semiconductors and, therefore, opens new avenues for revolutionary applications, such as RF devices [8], sensors [9], and high-precision metrology [10].

One of the earliest efforts of producing monolayer graphite (graphene) was performed by Lang et al. in 1975 [11]. They produced mono- and multi-layered graphite by thermal decomposition of carbon on single crystal Pt substrates. After graphene discovery, several techniques have been developed to fabricate monolayer (ML) or multilayer graphene. As mentioned, one of the methods is mechanical exfoliation of graphite, which is basically a repeated peeling process. In 2004, Andre Geim and Konstantin Novoselov extracted single graphene sheet by rubbing or by scratching the graphite surface with sticky tape until they thin it to a single flake of graphite, graphene. Unfortunately, the largest flakes obtained by this way are typically about 10–100 μm, which is only useful for fundamental laboratory research but irrelevant for industrial purposes [4]. Otherwise, chemical exfoliation can be introduced as a two-stage process: The first stage is increasing the space between layers in graphite, which results in reducing the interlayer van der Waals forces. This can be done by intercalating graphite to make graphite-intercalated compounds (GICs) [12]. As the second stage, these GICs can be exfoliated to a single- or few-layer graphene by ultra-sonication or rapid heating. This process is a promising method to synthesize large-scale graphene, but the structure has a lot of defects due to the oxidation and reduction chemical processing. An alternative technique is chemical vapor deposition (CVD) of graphene on transition metal substrates, such as Cu, Ni, Pd, Au, or Ru [13, 14]. As some of transition metals can be etched by acid solutions, graphene on these materials can be transferred to other substrates. The first report on few-layer graphene synthesized by CVD appeared in 2006 [15]. This method is based on the saturation of transition metal by carbon in a hydrocarbon gas ambient at high temperature. In this process, metal substrate works as a catalyst of reaction, and variables such as C solubility determine deposition, the exact mechanism and conditions of graphene formation, which additionally has an effect on graphene quality. The advantage of this method is that large graphene domains can be obtained, but this technique requires transfer to an insulating substrate, with methods that have yet to be developed and optimized to preserve pristine graphene characteristics.

In comparison with other methods, the technique that has been developed for simple (one-step) fabrication of large-area, low-defect density graphene films directly on semiconducting substrate is based on high-temperature decomposition of SiC by "preferential" atomic Si sublimation [8, 16–21]. The basic principles of a modified seeded sublimation growth process for the growth of 6H-SiC were established in 1970 by Tairov and Tzvetkov [22, 23]. This process, also referred to as the modified Lely process, was a breakthrough for SiC growth. In 1975, van Bommel et al. described the epitaxial sublimation of silicon from single crystals of SiC(0001). At elevated temperatures under ultrahigh vacuum (UHV; $<10^{-10}$ Torr), mono-layered flakes of carbon consistent with the structure of graphene were obtained, as determined by low-energy electron diffraction (LEED) and Auger electron spectroscopy [2].

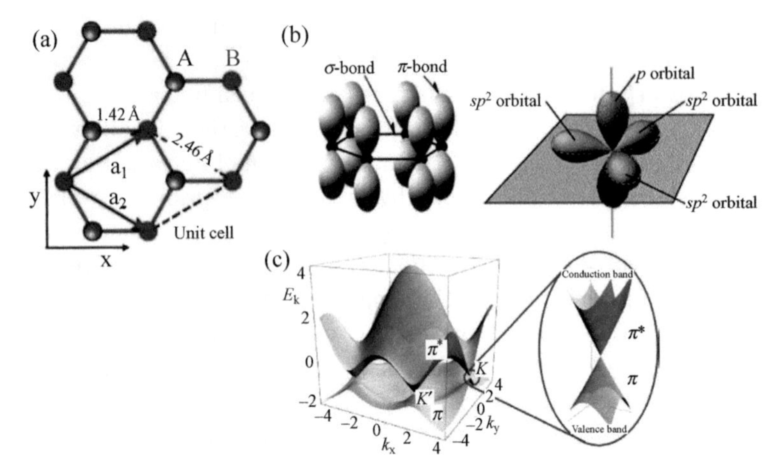

Figure 3.1 (a) Bravais lattice of the graphene; (b) σ- and π-bonds in graphene; (c) graphene π- and π*-band structure. Reprinted from Ref. [7], with permission from American Physical Society.

A simplified explanation is that graphene growth by sublimation method is performed in a furnace with an Ar overpressure, which improves the uniformity of the epitaxial graphene (EG) layer [21]. The main advantages of EG growth on SiC are that no transfer is needed for electronic device fabrication and the size of the graphene sheet can be, in principle, as large as the substrate, which is another benefit for device processing. As a polar material, basal-plane SiC wafers have two inequivalent terminations: the one called Si-face corresponding to the (0001) polar surface and the C-face (000-1). For both the Si-face and C-face, the growth mechanism of graphene layers is driven by the same decomposition process: sublimation of Si atoms at elevated temperatures at a rate much faster than C atoms due to its higher vapor pressure [24]. The remaining C forms a graphene film on the surface. However, the surface reconstructions and growth kinetics for Si- and C-faces are different, resulting in different graphene growth rates, growth morphologies, and electronic properties [2, 25]. The underlying SiC substrate, having the space group $P6_3mc$ with a hexagonal lattice, provides excellent symmetry matching for graphene epitaxy, which is why graphene grown in this manner is referred to as "epitaxial graphene" [26]. EG on SiC is a promising method for quality graphene growth and shows interesting physical characteristics such as high-frequency transistor behavior [8, 27], ballistic transport in nanoribbons [28], metrology standard quantum Hall effect [10], and half-eV bandgap structures [29].

3.2 Epitaxial Graphene on SiC Polytypes

Growth of EG is based on annealing the SiC crystal at high temperature. As silicon has a higher vapor pressure than carbon in the SiC substrate [30], the Si atoms desorb first from the sample surface during annealing process and leaving the C atoms behind, leading to a carbon-rich surface to emerge, until final graphene is formed. In 2004, de Heer et al. utilized this process for the first time, on purpose, to produce graphene layers [26], although surface graphitization of SiC had been observed before [31]. The process was first performed in vacuum, not in an Ar atmosphere, but nowadays growth is typically performed in an Ar overpressure, as mentioned, to improve the uniformity of the EG film [21].

Graphene is a material of atomic thickness, which implies that the characterization methods should meet certain requirements, e.g., in terms of resolution and source energy. The most common characterization techniques for graphene layers are low-energy and photoemission electron microscopy (LEEM and PEEM), Raman spectroscopy and microscopy, scanning tunneling microscopy (STM), atomic and electrostatic force microscopy (AFM and EFM), transmission electron microscopy (TEM), and scanning electron microscopy (SEM). These characterization techniques have been introduced in detail (especially for graphene) in Ref. [32].

3.2.1 SiC as a Substrate

The first SiC was synthesized in 1824 by the Swedish scientist Jöns Jocab Berzelius [33]. In the following decades, only non-systematic studies of SiC synthesis in small quantities on a laboratory scale have been reported. The process of SiC powder production was introduced in 1892 by E. G. Acheson [34]. The first SiC production on an industrial scale dates back to 1893 by Acheson [35]. In this process, SiC was produced by the electrochemical reaction of sand and carbon at high temperatures (up to 2550°C). Because of its extreme hardness, the resulting material was used in polishing applications. The electroluminescence of SiC light-emitting diodes, the first electrical property of SiC, was measured in 1907 [36]. As production of high-quality SiC crystals is necessary for application in electronics, in 1955 Lely introduced a new method for the growth of high-quality SiC, which was based on sublimation and enabled growth of SiC platelets [37]. Tairov and Tsvetkov improved this process in 1978 when they introduced an SiC seed crystal on which the vapor species deposited, resulting in a boule of the material [22]. This method is called sublimation growth and is based on physical vapor transport (PVT). Nowadays, there are commercially available single crystal wafers of 4H-SiC with micropipe densities less than 1 cm^{-2} in 6-inch wafers. A common method to grow SiC epitaxial layers is CVD; the advantage of this method is in providing good structural quality and excellent doping control.

Needless to say, SiC never reached the importance that Si has for (electronic) device technology, which can mainly be attributed to difficulties in the crystal growth. However, SiC has found its place in some specific areas in the markets, where Si reaches its limits. For example, SiC is used in high-power and high-frequency as well as high-temperature device applications. Since high-quality EG can be formed on SiC surface by thermal decomposition of SiC [21, 26], an additional future commercial application for SiC could be the epitaxial growth of graphene on SiC wafers.

SiC is a IV–IV compound semiconductor with mainly covalent Si–C bonds (88% covalent and 12% ionic). SiC is the only chemically stable compound containing only Si and C atoms. Its crystalline structure consists of close-packed stacking of double layers of Si and C atoms. The fundamental unit in the SiC structure is a covalently bonded tetrahedron with fourfold symmetry, consisting of either SiC4 or CSi4, as shown in Fig. 3.2. The distance between the two neighboring silicon or carbon atoms, **a**, is about 3.08 Å, while the very strong sp^3 bond between carbon and silicon atoms, **b**, relates to their very short distance, approximately 1.89 Å.

As shown in Fig. 3.2b, the spacing between the silicon layers is approximately 2.51 Å. The unit cell is bonded through the corner atoms of the tetrahedron; as shown in Fig. 3.2b, there are two possible orientations of adjacent tetrahedral (by 60° rotation). The various rotations and translations conduce to the many different stacking arrangements (or polytypes) of the Si–C bilayers along the *c*-axis.

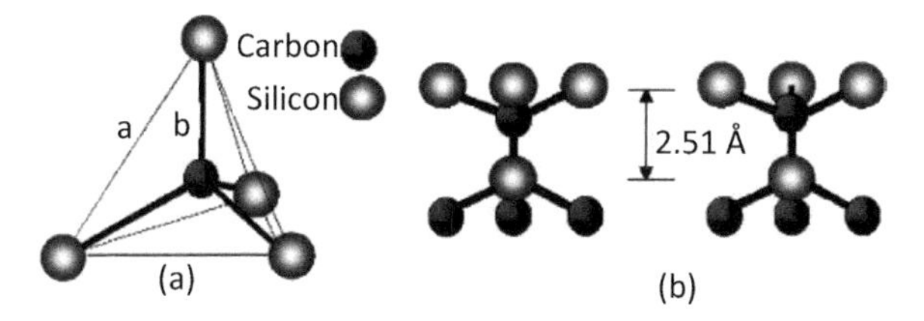

Figure 3.2 (a) Basic unit cell of silicon carbide. The distance between Si–Si or C–C atoms, **a**, is about 3.08 Å, and that between C–Si atoms, **b**, is approximately 1.89 Å. (b) The two configurations of silicon and carbon atoms, rotated 180° [32].

The possible atomic arrangements of the atoms in the hexagonal wurtzite unit cell are shown in Fig. 3.3. One can denote the first layer of atoms with position A, and then the atoms in the next layer may sit at either position B or position C. Thus, the simplest polytype is the 2H (ABAB...). The cubic structure (zincblende) of 3C-SiC has a stacking sequence of ABCABC... (or ACBACB...). The most common forms of SiC are the 6H and 4H polytypes with the ABCACBABCACB... and the ABCBABCB ... stacking, respectively. Here we use Ramsdell notation, which is common to describing the polytypes [38]. The number in the name of polytype refers to the number of layers needed to repeat the pattern and the letter in a polytype's name corresponds to the first letter of the crystal system (C for cubic, H for hexagonal, and R for rhombohedral).

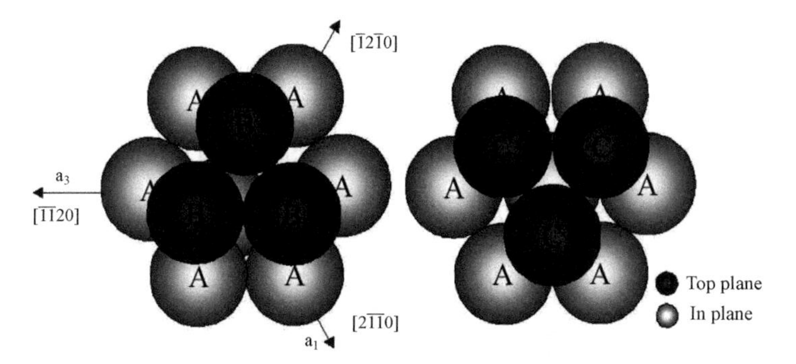

Figure 3.3 Possible stacking orientations of atoms in a close-packed hexagonal structure [32].

Figure 3.4 shows the stacking sequence for 3C-, 4H-, and 6H-SiC polytypes. Since there is no rotation in the stacking sequence of cubic polytype compared with the hexagonal polytypes, the 3C structure proceeds in a straight line and hexagonal structures proceed in a zigzag pattern. The A position in 4H-SiC is a cubic site, and B position is a hexagonal site. In 6H-SiC, the A position is a hexagonal site, and B and C are cubic. More than 200 SiC polytypes have been found, some with a stacking period of several hundred bilayers [39]. Importantly, the properties of SiC, such as bandgap, depend on the polytypes and also the atom position and its surrounding in the polytype.

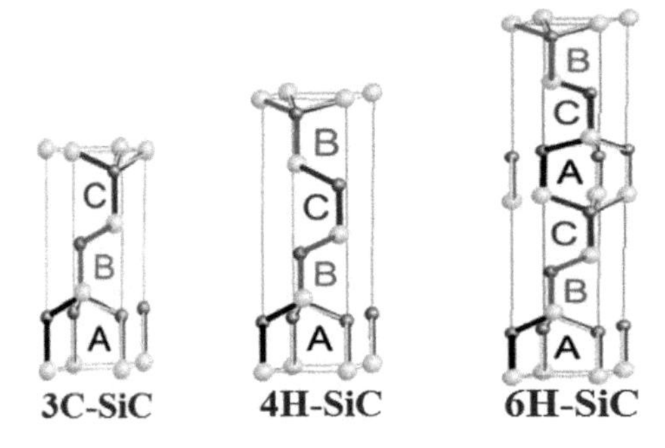

Figure 3.4 Stacking sequence of SiC polytypes. Reprinted from Ref. [17], with permission from Elsevier.

Both wurtzite and zincblende structures have polar axes due to the lack of inversion symmetry. The polarity of SiC can be defined with respect to the position of the Si atom in the (0001) bilayer. In the Si-face of SiC, the Si atom occupies the top position in the bilayer, while in the C-face of SiC, the top position is occupied by a C atom (Fig. 3.5).

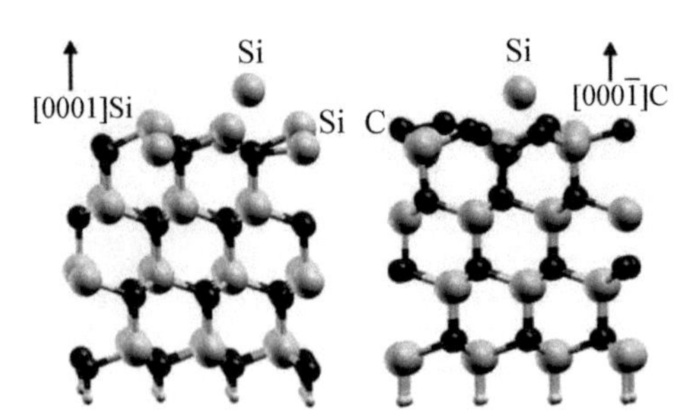

Figure 3.5 Schematic of the SiC polar structures: left image shows Si-face, and right image shows C-face [32].

3.2.2 Graphitization Process of SiC Polytypes

Growth of graphene on SiC surfaces, which is in principle a graphitization process under controlled conditions, can be executed in several experimental setups. Here we show the example of a vertical RF-heated furnace consisting of a quartz tube, porous graphite insulation, and graphite crucible (Fig. 3.6).

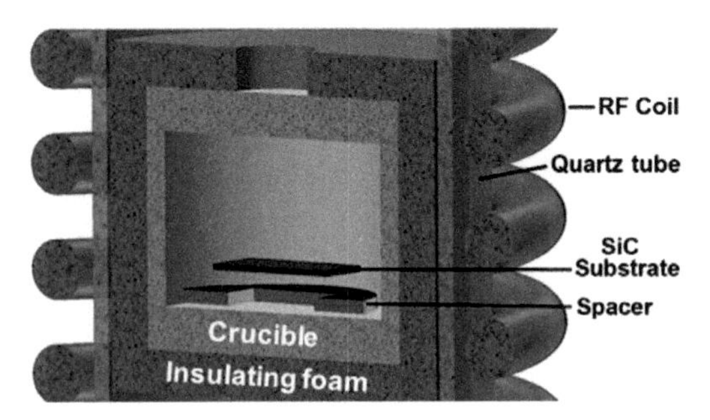

Figure 3.6 Sketch of a vertical RF-heated furnace [32].

The growth mechanism of EG for both the Si-face and C-face graphene layers is driven by sublimation of Si at high temperatures with a rate faster than C due to its higher vapor pressure (more details in the following). The remaining C tends to form a graphene film on the surface. However, the SiC surface reconstructions and growth kinetics for each polar surface are different, which results in different graphene growth rates and morphologies as well as electronic properties [25, 40–42].

As illustrated in Fig. 3.7a, the 4H-SiC polytype has two kinds of decomposition energies, terraces 4H1 (−2.34 meV) and 4H2 (6.56 meV), respectively, and the 6H-SiC (Fig. 3.7b) has three distinct terraces −6H1 (−1.33 meV), 6H2 (6.56 meV), and 6H3 (2.34 meV), while 3C-SiC (Fig. 3.7c) has only one kind of terrace 3C1 (−1.33 meV) [43].

The growth process is schematically showed for 4H-SiC in Fig. 3.8. The growth of the EG is not uniformly distributed over the SiC substrate surface. Since the Si and C atoms bonded more weakly in the vicinity of step edges, Si desorbs from these areas faster in comparison with the terraces. It is worth noting that C contained in one unit cell (three Si–C bilayers) of SiC substrate, in principle, would be sufficient to allow the formation of one-layer graphene.

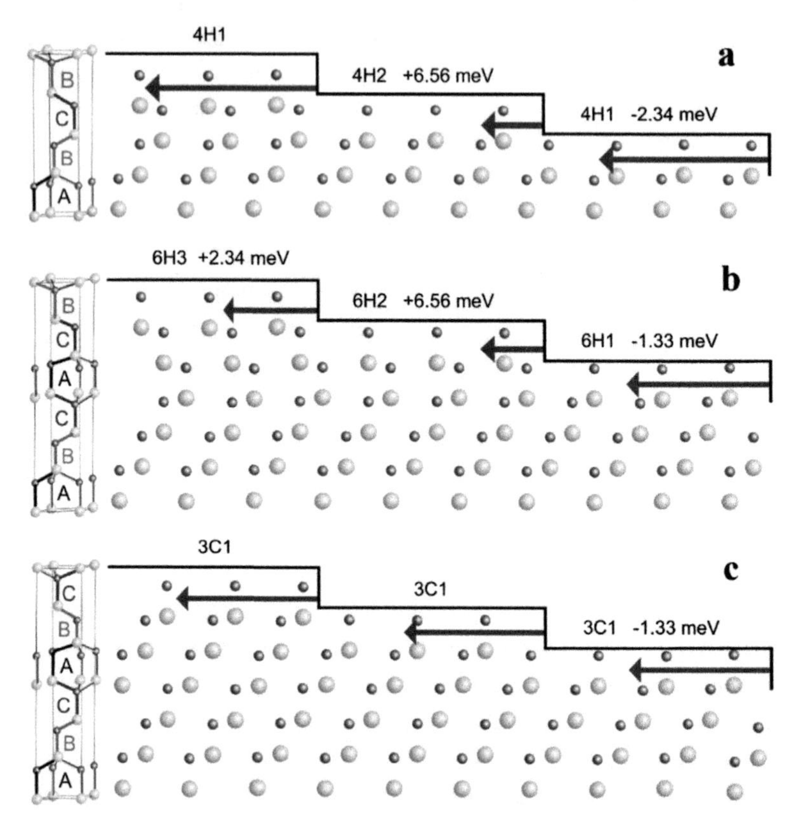

Figure 3.7 Stacking sequences and possible terraces on (a) 4H-SiC, (b) 6H-SiC, and (c) on 3C-SiC surfaces. Large and small circles represent Si and C atoms, respectively. Length of arrows indicates different step-decomposition velocities. The surface energies needed to remove a particular terrace are specified. Reprinted from Ref. [17], with permission from Elsevier.

Based on the mentioned terrace energies for 4H-SiC, it will cost less energy to remove a 4H1 terrace; in other words, for the 4H1 terrace, the step-decomposition velocity will be faster (Fig. 3.7a). As shown in Fig. 3.8, from the edge of the 4H1 terrace, free surface C atoms are emitted onto the terrace as Si atoms leave the surface (stage 1). The C atoms coalesce and nucleate into graphene islands (stages 1 and 2), which in turn act as sink sites for subsequently emitted C atoms [44]. After the 4H1 terrace step catches the 4H2 step, the newly formed two SiC bilayer-height step provides more C atoms as compared to the one bilayer-height step and the first graphene layer extends along the step edge (stage 2). The large percentage of bunched steps with four Si–C bilayers, i.e., an increased source of carbon, will impose the formation of a second layer graphene (Fig. 3.8, stage 3) since some extra C will be released. Therefore, a full coverage of the 4H-SiC substrate surface by just one-layer graphene may be challenging.

A similar mechanism of energy minimization is expected in the 6H-SiC polytype (Fig. 3.7b). As a result, first the step 6H1 will catch step 6H2 and form two Si–C bilayers. Then step 6H3 will advance and merge with the two bilayer step. The growth process for 6H-SiC is the same as for 4H-SiC, which was explained above. Most outstandingly, significant step bunching occurs; the initial steps flow over the surface while graphene is forming, bunch together, and form higher steps with larger terraces in between [17, 45–47]. Differently, on 3C-SiC all terraces have the same decomposition energy (Fig. 3.7c) and no energetically driven step bunching should be expected. Yet, in this polytype, a non-uniform sublimation can be induced by the presence of extended defects such as stacking faults, which are characteristic of this material (see Section 3.2.4).

Structural and electronic properties of graphene grown on SiC are strongly affected by the substrate polytype and polarity. As mentioned above, the surface reconstructions and growth kinetics for Si- and C-faces of the SiC substrate surface are different. The first layer graphene grown on the Si-face of SiC polytypes is known as a zero layer or buffer layer, which includes sp^3 bonds of C atoms to Si atoms of substrates. The next layer of graphene grown on the first layer, known as a monolayer (ML) graphene, and C atoms just have sp^2 bonds with C atoms around (inset left side in Fig. 3.8) and week van der Waals with respect to the buffer layer. In contrast, for the graphene deposition onto SiC C-face, the buffer layer has not been reported. In the following sections, we will focus on the differences of growth mechanism, structural properties, and electrical properties on 4H-, 6H-, and 3C-SiC polar faces.

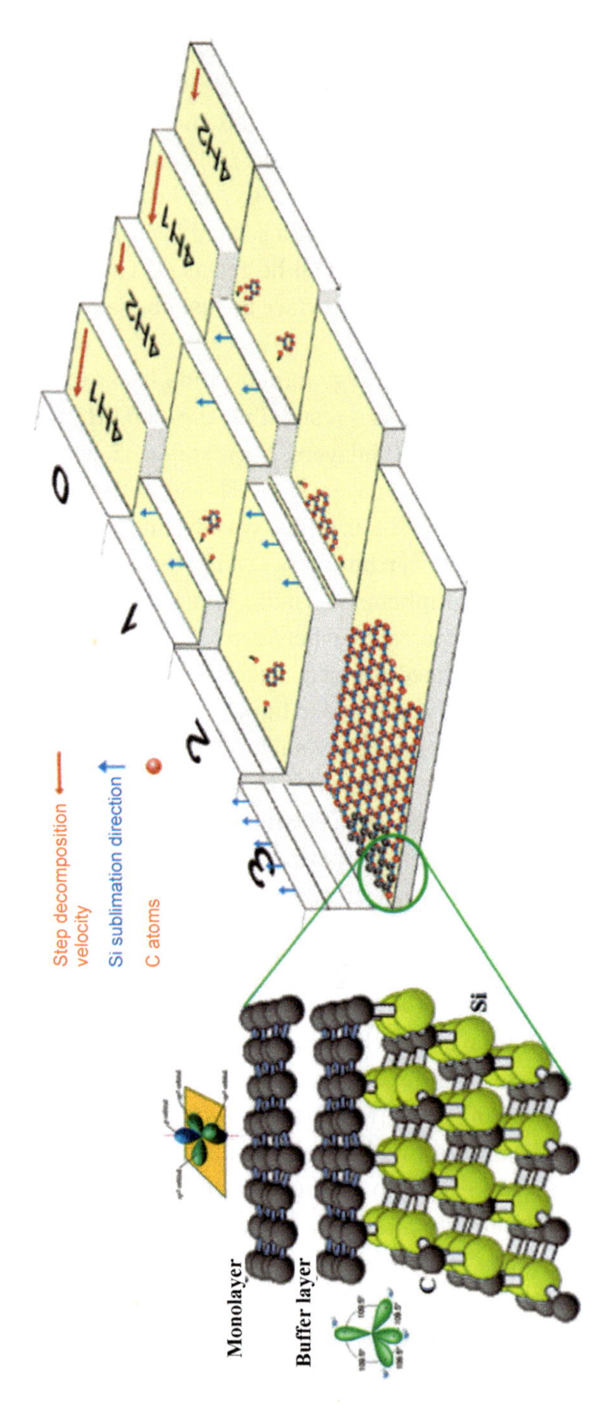

Figure 3.8 Schematic depiction of the growth process of epitaxial graphene via sublimation of Si from 4H-SiC surfaces. Reprinted from Ref. [17], with permission from Elsevier.

3.2.3 Growth of Graphene on SiC Polar Faces

The most commonly used SiC polytype structures for growing EG are 4H-SiC and 6H-SiC. Both of these SiC polytypes have either an Si- or a C-terminated surface. There is an epitaxial degree for these facets and the graphene hexagonal lattice.

In terms of history of EG on SiC, the first experimental works with respect to graphitization of SiC crystals at temperatures about 2050–2280°C in vacuum (10^{-5} Torr) dates back to 1962 by D. V. Badami [52]. He studied the samples with X-ray diffraction and found that the *c*-axis of graphite is along the *c*-axis of the hexagonal SiC crystal [53]. In 1975, van Bommel et al. showed that the graphite structure was formed on a hexagonal SiC(0001) surface via a $(6\sqrt{3} \times 6\sqrt{3})R30^\circ$ structure by LEED experiments [2]. He found that the carbon formation rate was greater on the SiC(000-1) polar face than on the SiC(0001) face. In 1998, Forbeaux et al. studied the evolution of SiC surface as it became more graphite-like (Fig. 3.9) by LEED and angle-resolved photoemission spectroscopy (ARPES) measurements [31]. They showed that with the increase in annealing temperatures, the concentration of Si vacancies in the top layers increased, so that there was larger number of C atoms surrounded by Si vacancies. In 2002, Charrier et al. showed that the thermal decomposition of 6H-SiC after annealing at increasing temperatures between 1080 and 1320°C led to the layer-by-layer growth of unconstrained, heteroepitaxial single-crystalline graphite by using grazing-incidence X-ray diffraction and STM. They could control the Si sublimation rate to form single- or a few-layer graphene on 6H-SiC(0001) [54]. In 2004, the same year when Novoselov et al. reported the electric field effect on mechanically exfoliated graphene [4], de Heer's group reported on their highly ordered graphene samples grown in vacuum, which showed Shubnikov–de Haas oscillations that corresponded to nonlinearities in the Hall resistance, indicating a potential new quantum Hall system. They also showed the Dirac nature of the charge carriers and exceeding the carrier mobility of 25,000 $cm^2/V\,s$ in graphene on SiC [18, 26]. In all the aforementioned research works, the growth of EG on SiC was done in UHV conditions, which resulted in an inhomogeneous graphene thickness. In 2008, Yakimova's group developed a novel method for homogeneous graphene growth by annealing SiC in an argon atmosphere, where Ar gas suppressed the excess decomposition and improved the quality of graphene [48]. In 2013, the same group studied the growth of graphene on 3C-, 6H-, and 4H-SiC and, for the first time, established the influence on the graphene formation of crystal structure at an atomic level and for each polytype [17]. The growth of graphene on SiC has kept progressing, and recently large-area homogeneous monolayer, bilayer, and few graphene layers on SiC(0001) and SiC(000-1) can be obtained.

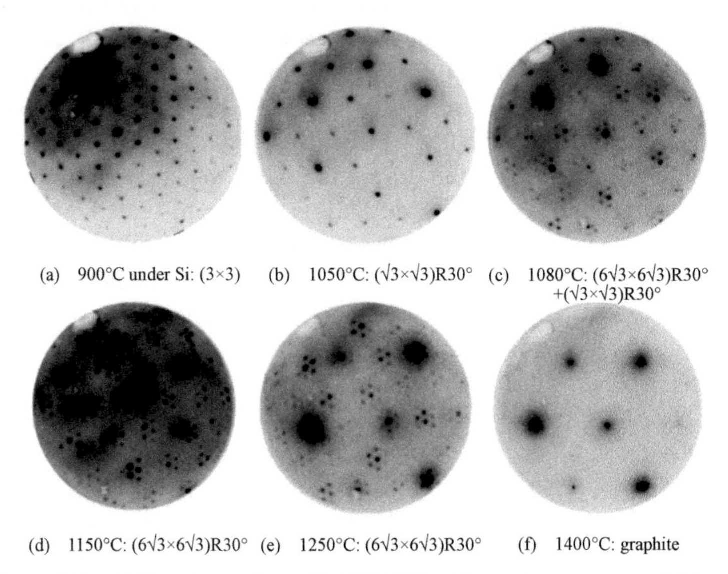

Figure 3.9 LEED patterns from 6H-SiC(0001) with a primary energy of 130 eV obtained on (a) (1×1), (b) $(\sqrt{3}\times\sqrt{3})R30°$, (c) $(\sqrt{3}\times\sqrt{3})R30° + (6\sqrt{3}\times6\sqrt{3})R30°$, (d, e) $(6\sqrt{3}\times6\sqrt{3})R30°$, and (f) graphite (1×1). Reprinted from Ref. [31], with permission from American Physical Society.

3.2.3.1 Effect of ambient conditions

The ambient conditions have a strong influence on the graphene quality in both polar faces of SiC polytypes. For example, the growth of EG on SiC(0001) in UHV yields graphene layers with low quality and small grains [55, 56]. At high temperatures, silicon atoms leave the surface at a high sublimation rate, and some carbon atoms stay on the surface. This is a process far from equilibrium, which leads to SiC substrate roughening. It has been shown that more homogeneous graphene layer can be produced by using short-time high-temperature annealing, in comparison with longer time heating at lower temperature [57, 58]. By using higher annealing temperature, the kinetic energy and mobility of C and Si atoms will be higher, which makes easier the surface restructuring. But this should be completed before graphene layer is formed, i.e., to avoid complete/massive graphitization, the Si sublimation should be suppressed. This can be done by having a control upon Si background pressure, and there are several methods to reach this goal.

Yakimova's group, in 2008, developed a novel method for EG growth on SiC substrates. The method is based on high-temperature (2000°C) sublimation at 1 atm of Ar ambient, which yields ML graphene on a large area (up to 20 × 20 mm^2) and allows good control over the thickness uniformity [16, 48]. The role of Ar is to suppress too high Si atoms sublimation and to make the process of Si depletion more uniform. Figure 3.10 shows LEEM images of the graphene layers grown in vacuum and Ar ambient. These results show an obviously better thickness uniformity and quality for graphene grown in Ar pressure [48].

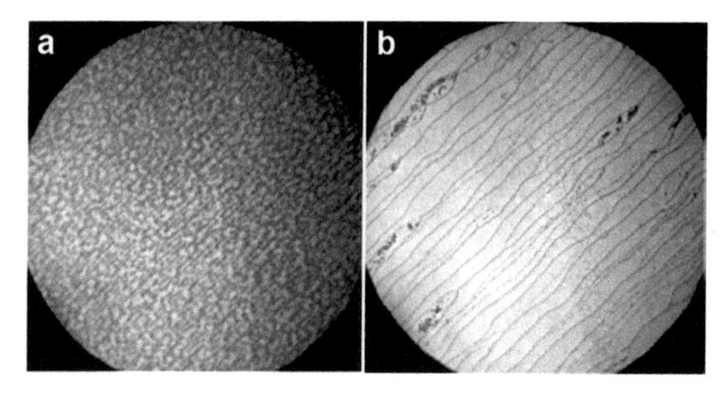

Figure 3.10 Graphene layers grown in (a) vacuum and (b) Ar. Reprinted from Ref. [48], with permission from American Physical Society.

In 2009, Emtsev et al. also reported Ar buffer gas around 1000 mbar to suppress the evaporation of Si atoms [21]. This allowed them to use temperature at around 1650°C, and no pit formation was observed when the terraces retract upon Si sublimation. Transport measurements of the graphene layer indicated higher motilities (2000 cm^2/V s) in comparison to UHV-grown EG (for 700 cm^2/V s) [49]. Tromp, Hannon, and other researchers controlled the Si background pressure using disilane gas [51, 59]. The presence of extra Si flux allows the surface to reach equilibrium with a pressure–

temperature phase diagram. An alternative is de Heer et al.'s work; by the so-called confinement controlled sublimation (CCS) method, a confined cavity could retain a finite Si background pressure as Si evaporates from SiC substrate [50]. The near-equilibrium CCS method has been demonstrated to be suitable to produce high-quality uniform graphene layers on both the Si-face and the C-face of 4H- and 6H-SiC. Further control of the graphitization rates can be obtain in this method by introducing inert gasses, e.g., to inhibit the graphene growth even at temperatures exceeding 1600°C [50].

The growth of EG and its final structure, growth morphologies, and electronic properties are strongly dependent on which SiC polar faces are initially exposed. For homogeneous growth of graphene, Si-face is a better choice. In the following, we will focus on the characteristics of the growth of graphene in Ar ambient on these two polar faces.

3.2.3.2 Growth of graphene on Si-face

The most outstanding advantage of graphene on the Si-face is that it is easier to control the thickness of graphene at the wafer scale. This control can be reached by optimizing the growth temperature and Ar pressure. The growth of ML graphene on the Si-face occurs via a step-flow process, as shown in Fig. 3.8. Reviewing, on this face Si sublimation initially results in a C-rich $(6\sqrt{3}\times 6\sqrt{3})R30^\circ$ structure (shortly $6\sqrt{3}$) nucleates at the step edges, which has the same honeycomb structure of graphene and known as the buffer-layer or zero-layer graphene [20, 48, 57, 60–62]. This reconstructed surface loses its distinct graphene electronic properties because of sp^3 hybridization of about 30% carbon atoms and covalently bonded with the Si atoms of the SiC interface (left side inset in Fig. 3.8) [48, 63, 64]. Because of this strong coupling with the substrate, buffer layer does not show graphene-like π bands and is electronically inactive [65]. After the formation of the buffer layer, further heating leads to the decomposition of SiC bilayers underneath the buffer layer, leading to the possibility of nucleation of a new, freestanding carbon layer (left side inset in Fig. 3.8) [66].

TEM and STM studies showed that buffer layer is located at 1.97 Å [67], 2.0 Å [68], 2.3 Å [69], or 2.5 Å [60] distance from Si-terminated SiC surface, which is less than the graphite inter-plane spacing of 3.35 Å. Differently, the spacing between the first and the second graphene layers is about 3.5 Å. Graphene layers on the Si-face are stacked with a 30° rotation, and the orientation relation is $[1\ \bar{1}\ 0\ 0]_{SiC} \| [1\ 1\ \bar{2}\ 0]_{graphene}$ and $(0\ 0\ 0\ 1)_{SiC} \| (0\ 0\ 0\ 2)_{graphene}$ [31]. From experimental and theoretical studies, it is recognized that the multilayer graphene electronic properties are strongly dependent on the stacking sequence.

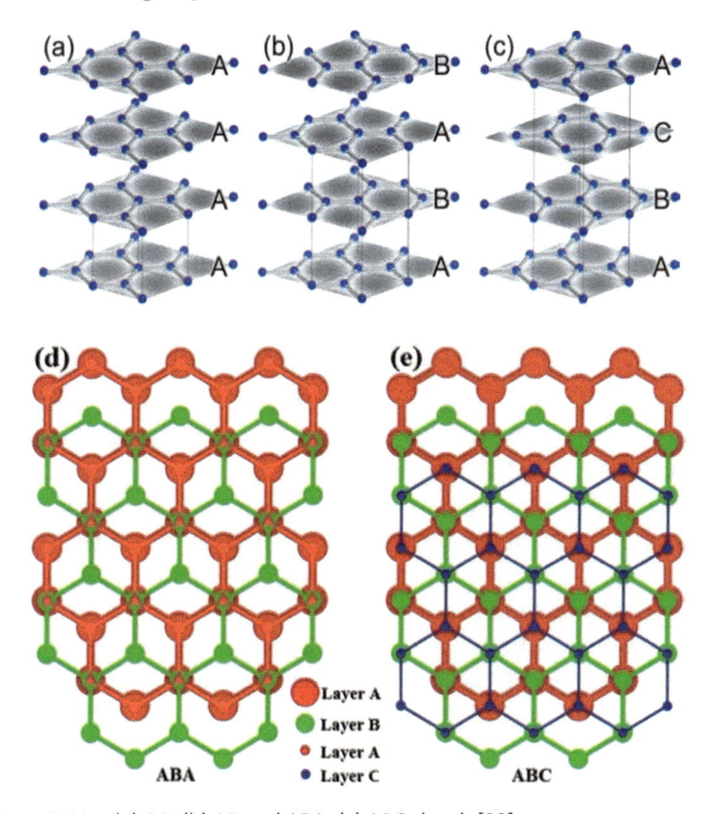

Figure 3.11 (a) AA, (b) AB and ABA, (c) ABC sketch [32].

For bilayer graphene, Bernal AB-stacking graphene, which is more stable than AA-stacking, shows a metallic behavior (zero gap) with chiral parabolic dispersions near the K point [70]. In AA-stacking, all carbon atoms are above each other, but in a Bernal structure, carbon atoms in the layer B are directly above the center of a carbon hexagon in the layer A (Fig. 3.11a,b). For trilayer graphene, two stable crystallographic configurations, which are energetically close, are ABA and ABC (rhombohedral) stacking orders (Fig. 3.11b–e). In a rhomboheral structure, the center of a carbon hexagon in the layer A is directly below a corner of a hexagon in the layer B, which is in turn directly below a nonequivalent corner of a hexagon in the layer C (Fig. 3.11c,e). For graphene tetra layer, four crystallographic configurations are predicted: ABAB, ABCA, ABAC, and ABCB stacking order.

In our early results [48], we have shown that large and homogeneous areas of single-layer graphene on top of an on-axis cut SiC(0001) substrate can be obtained in argon ambient. Using a high temperature of about 2000°C results in an increase in the sublimation rate of Si atoms, surface mobility of the carbons, and nucleation rate of the buffer layer. However, using Ar ambient instead of vacuum suppresses a too fast decomposition rate and provides a smoother decomposition of the SiC. As a result, formation of larger terraces with single-layer graphene films on top is obtained, as shown in Fig. 3.10.

The corresponding TR-CAFM current map and torsional morphology of graphene grown (at 2000°C and in Ar ambient) on-axis 6H-SiC(0001) and 8° off-axis 4H-SiC(0001) are shown in Fig. 3.12 and Fig. 3.13, respectively. Comparing the morphology in these two images shows that while on the off-axis cut sample, a significant step bunching was observed, a flatter surface (RMS = 2.4 nm) is obtained on the on-axis cut sample. The current maps are also distinctly different. A comparison of Fig. 3.12b and Fig. 3.13b shows that graphene layer on an on-axis cut sample shows a uniform current all over the scanned area, which suggests nanoscale uniformity [71]. The bright lines in the topography signal correspond to the wrinkles on (epitaxial) graphene [72]. They form due to the thermal expansion mismatch between graphene and SiC during the sample cooling, and they have characteristic higher current density.

The off-cut angle affects the growth of graphene with multilayer graphene as it provides larger number of step edges on which C nucleates. For instance, few layers of graphene grown on 8° off-axis 4H-SiC(0001) substrates grown in Ar ambient with a temperature range of 1600–2000°C have been characterized by AFM and HR-XTEM (Fig. 3.14) [73]. The results show that FLG covers the 100–200 nm wide terraces of the SiC surface for all the growth temperatures. The wrinkles, 1–2 nm high and 10–20 nm wide, are preferentially oriented in the direction perpendicular to the step edges. This parallel orientation of wrinkles appears on the off-axis SiC surface, while for the EG grown on the on-axis 4H-SiC, an isotropic mesh-like network of wrinkles forms. The observed phenomenon deserves further investigation.

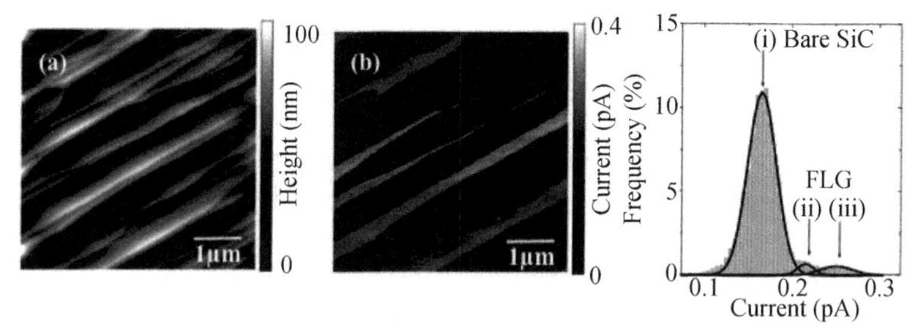

Figure 3.12 AFM (a) height image, (b) conductive map, (c) histograms extracted from the current map in (b). Evaluated three regions (region with graphene coverage (curves (ii) and (iii)) and region devoid of graphene (curve (i)) are indicated distinctly. Reprinted from Ref. [71], with permission from Trans Tech Publications.

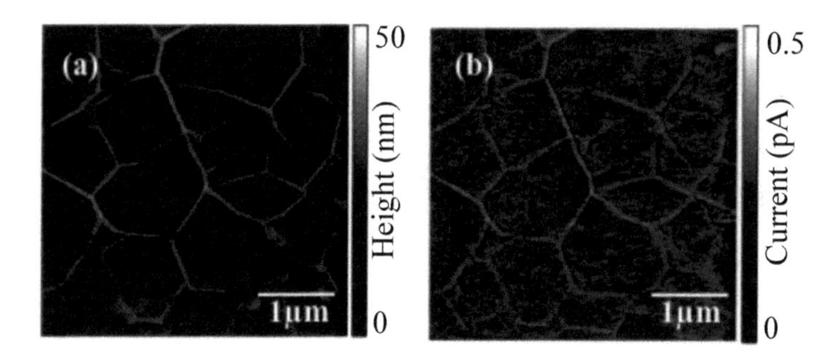

Figure 3.13 TR morphology (a) and current map (b) collected on an on-axis sample. Reprinted from Ref. [71], with permission from Trans Tech Publications.

Berger et al. showed that the buffer layer exhibits a large bandgap and a Fermi level pinned by a state having a small dispersion that relates to the dangling bonds in between the bulk SiC and this buffer layer [74]. The existence of this buffer layer is an obstacle for the development of future electronic devices from graphene grown on the Si of SiC, because it may affect its transport properties. However, hydrogen, sodium [75, 76], oxygen [77], lithium [78], Si [79], gold [80], fluorine [81], and germanium [82] intercalation at high temperatures can transform the buffer layer into a graphene layer with the consequent improvement in the electrical properties [83]. We have used different elements for intercalation, such as H, Li, Na, and Si [75, 76, 78, 79]. Hydrogen intercalation studies by ARPES and energy dispersive XPEEM measurements confirm that the buffer layer is converted into a graphene layer. It has been shown that H intercalation is a reversible process and the initial monolayer graphene plus carbon buffer layer situation can be recovered by annealing at a temperature about 950°C [75]. For Si intercalation, it is observed that Si is not able to penetrate through monolayer graphene when the sample is kept at room temperature. Intercalation can occur at an elevated temperature of about 800°C where the Si atoms tend to migrate preferentially through the graphene toward domain boundaries and likely other defect areas. For Na intercalation, the results show that partial intercalation occurs in between carbon layers, and at the interface occurred directly after deposition, but most of the Na initially remained on the surface, forming Na droplets. Intercalation was inhomogeneous and occurred on the ML and bilayer areas. Annealing at higher temperatures resulted in Na de-intercalation.

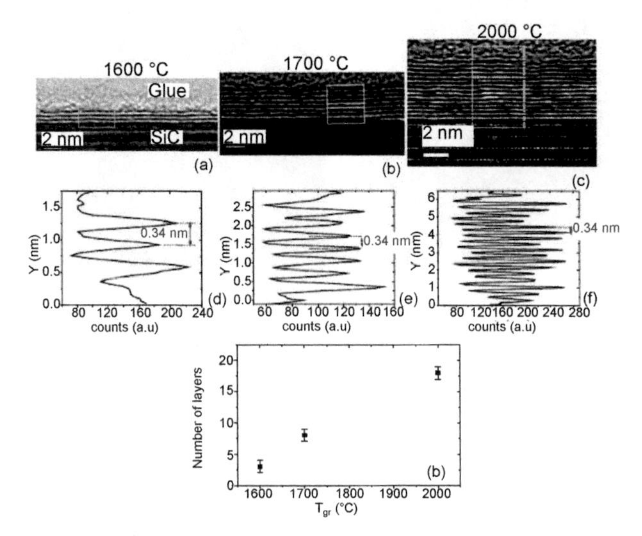

Figure 3.14 HRXTEM analyses of the samples annealed at different temperatures. Images of the samples annealed at 1600 (a), 1700 (b), and 2000°C (c), and corresponding linescans (d–f), showing 3, 8, and 18 layers grown on the surface of 4H-SiC, respectively [73].

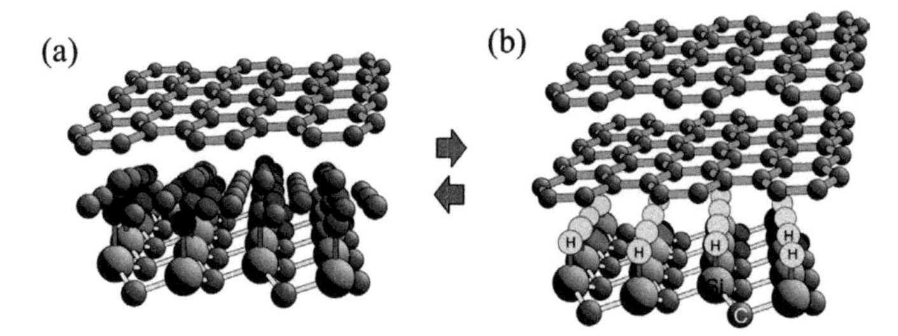

Figure 3.15 (a) A sketch of monolayer graphene on the SiC(0001) substrate, including a strongly bound buffer layer at the interface. (b) The structural model of new phase formed, having a hydrogen-intercalated layer in between the SiC substrate and the initial buffer layer, which has now become the second graphene layer. Reprinted from Ref. [75], with permission from Elsevier.

Surface conductance map (Fig. 3.16) have been done to study the local resistance enhancement due to characteristic features in EG grown on 4H-SiC(0001), such as substrate-related steps and the variation of the number of layers, and, in particular, the monolayer/bilayer junction [84]. The relationship between morphological and electrical maps revealed a local degradation of EG conductance specifically at the substrate steps or at the junction between ML and bilayer graphene regions. By conductance mapping on graphene layer coverage of 10 SiC steps with different heights, we found that the local resistance of ML graphene on steps decreases with decreasing step height. These step effect strongly depends on the charge transfer phenomena between the step sidewall and graphene, while increasing resistance at the ML/bilayer junction is a purely quantum-mechanical effect. The latter is due to the weak wavefunction coupling between the monolayer and bilayer bands, as demonstrated by the ab initio calculations.

The energy structure in the density of occupied states of graphene grown on n-type 6H-SiC(0001) has also been studied recently. As shown in Fig. 3.17, we create quantum well (QW) levels [85]. In this work, we showed that the structure in the valence band density of states near the Fermi level is described by the QW states whose number and energy position coincide with the calculated ones. The structure revealed with photoelectron spectroscopy is described by creation of the QW states whose number and the energy position ($E1 = 0.3$ eV, $E2 = 1.2$ eV, $E3 = 2.6$ eV) coincide with the calculated ones for deep ($V = 2.9$ eV) and narrow (width = 2.15 Å) QW formed by potential relief of the valence bands in the structure graphene/n-SiC. It can be assumed that the property is an attribute not only of graphene (or few-layer graphene) on the wide-bandgap semiconductor substrate but also of graphene on dielectric and of suspended graphene. The type of substrate can define the type of QW; n-type results in the formation of QW in occupied VB, whereas creation of QW in unoccupied VB is expected for p-type substrate. The QW state formation is due to large electron/hole mass in the direction perpendicular to the graphene plane.

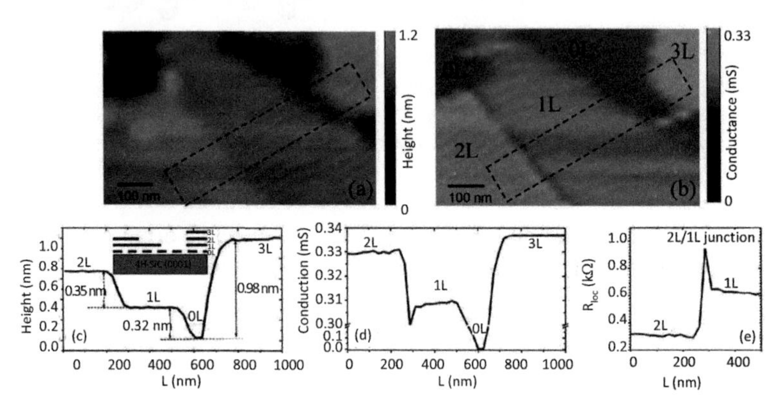

Figure 3.16 (a) AFM morphology and (b) conductance map on an epitaxial graphene with lateral variations in the number of layers. (c) Height profile extracted from the selected area on the morphology map, with the layer structure schematically represented in the insert. (d) Conductance profile extracted from the selected area in the conductance map. (e) Local sheet resistance of graphene calculated from the conductance profile in the region close to the monolayer/bilayer junction. Reprinted from Ref. [84], with permission from American Physical Society.

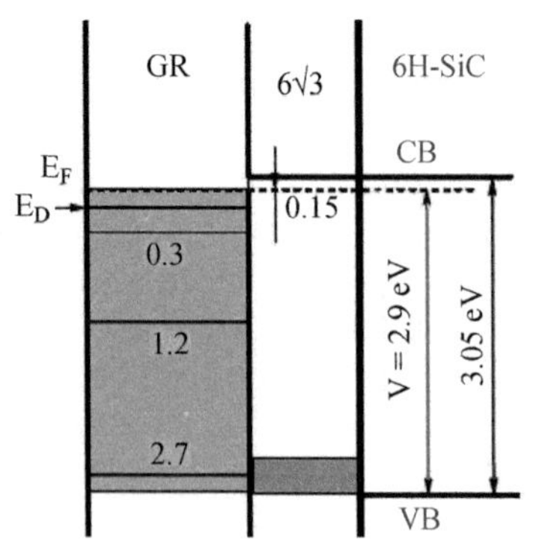

Figure 3.17 Scheme of the hole QW formed by graphene on SiC with $6\sqrt{3}$ interface carbon monolayer (buffer layer). Reprinted from Ref. [85], with permission from Elsevier.

3.2.3.3 Growth of graphene on C-face

The epitaxial growth of graphene on the C-face is faster and less controllable than on the Si-face. However, it seems to be relatively weakly attached to the underlying surface. No buffer layer has been detected, and the first graphene layer is located at a typical distance of about 3.2 Å from the SiC surface, which is too far to form covalent bonding between carbon atoms [86]. Yet LEED [87] and STM [88] results indicate that a significant degree of rotational disorder in the graphene films. Due to the rotational disorder, the electronic properties of the graphene on C-face become similar to freestanding monolayer graphene [89]. The band structure of each layer is linear and not parabolic as for bilayer graphene on the Si-face of SiC [90]. Therefore, using graphene layers grown on C-face we can improve the electronic quality of graphene on SiC electronic devices. On this face, graphene film likely grows following a 2 × 2 reconstruction and does not have an ordered interfacial layer that causes increased carrier scattering [2].

Our early experimental results of graphene growth on C-face of on-axis 4H-SiC substrates using a high temperature (1800–2000°C) in Ar ambient clearly showed the existence of distinct graphene grains related with the different azimuthal orientations (Fig. 3.18), showing that adjacent graphene layers are rotationally disordered. Our results indicate that the electron carrier concentration induced in the second and third graphene layers on the C-face is less than $\sim 4 \times 10^{11}$ cm^{-2} [91]. Rotational disorder was also understood from the sharp (1 × 1) μ-LEED patterns observed and that only six Dirac cones centered around the K-points in the Brillouin zone appeared in the constant energy photoelectron angular distribution patterns Ei(kx,ky) recorded from grains with 2, 3, and 4 MLs of graphene grown at lower temperature (1400–1500°C) [92].

Morphology, chemical composition, and surface potential of graphene grown on the C-face of 4H-SiC were also studied [93]. By matching the same nanoscale features on the surface potential and Raman spectroscopy maps (Fig. 3.19), individual domains and bare SiC substrate were assigned to graphene flakes of 1–5 monolayers in thickness. It is shown that the growth proceeds in an island-like fashion, consistent with the Volmer–Weber growth mode, rather than in a layer-by-layer manner as established on the Si-terminated face of SiC. Raman spectroscopy data show evidence of large-area crystallites (up to 620 nm) and high-quality graphene on the C-face of SiC. A comprehensive chemical analysis of the sample has been provided by X-ray photoelectron spectroscopy investigations, further supporting surface potential mapping observations on the thickness of graphene layers. It is shown that for the growth conditions used in this study, 5 monolayer thick graphene does not form a continuous layer, so even with such thickness, growth conditions are not optimized to completely cover the substrate. With the particular conditions used for graphene growth in the current study and considering that the Volmer–Weber mode is dominant, synthesizing large-scale monolayer graphene on the C-face of SiC remains a challenge. To benefit from the higher carrier mobility of graphene on the C-face compared to the Si-face, optimization of growth conditions by using different gas atmosphere is needed for enabling the growth of uniform ML graphene.

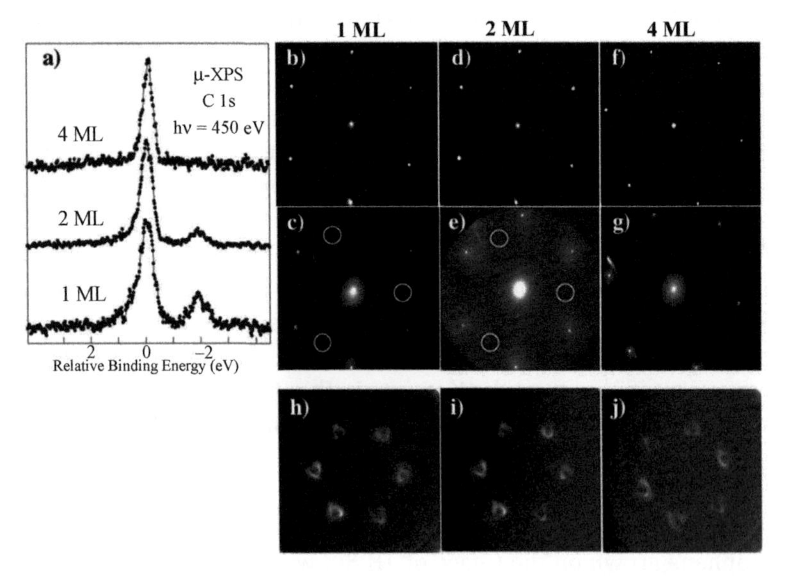

Figure 3.18 (a) C 1s spectra collected, using a probing area of 4 μm, from domains with different graphene thickness; (b)–(g) show 1.5 μm selected area LEED patterns from; (b) and (c), the 1 ML domain at 45 and 145 eV; (d) and (e), the 2 ML domain at 45 and 145 eV; and (f) and (g), the 4 ML domain at 45 and 130 eV, respectively. In (h)–(j), the photoelectron angular distribution patterns (PAD) collected from the 1 ML, 2 ML, and 4 ML domains, respectively, are shown. The PADs were collected at an energy of 1.5 eV below the Fermi level, using a probing area of 800 nm [92].

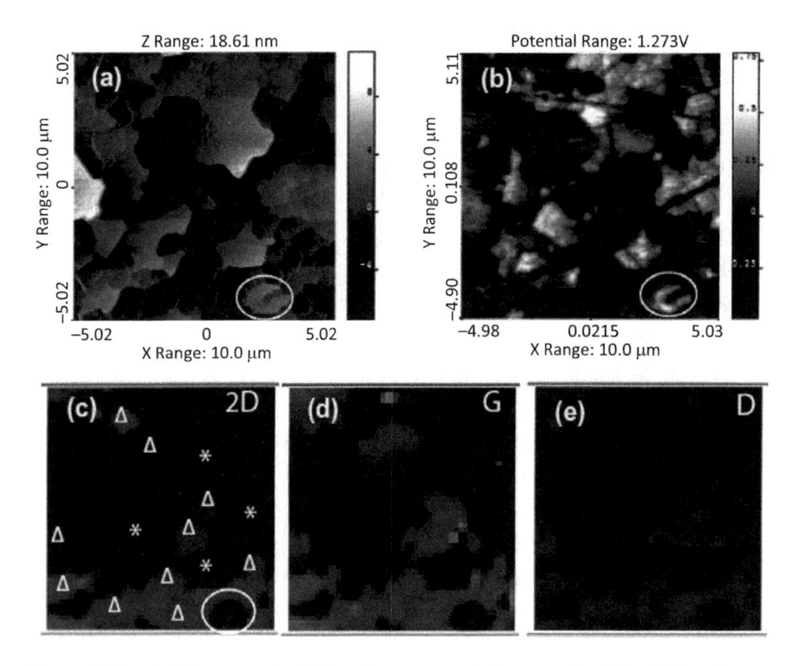

Figure 3.19 (a) Topography, (b) surface potential maps of a (10 × 10) μm^2 area of the sample. Spatially resolved Raman spectroscopy maps of the area depicted in (a) and (b) corresponding to 2D, G, and D peak intensities are shown in (c–e), respectively. As guide to the eye, A-like feature highlighted by the white circle in (a) and (b) can be correspondingly identified on the Raman spectroscopy maps. Reprinted from Ref. [93], with permission from Elsevier.

Recently, another study was done on the structural, vibrational, and dielectric function properties of graphene grown on C-face 4H-SiC at temperatures of 1800°C, 1900°C, and 2000°C [94]. We have shown that the average number of layers and the size of the domains with uniform thickness increase with the increase in process temperature. This improved graphene coverage and homogeneity are attributed to an enhanced sublimation of Si from the SiC. The improvement in graphene quality with the increase in growth temperature is further manifested by an increased crystallite size (a domain of solid-state matter that has the same structure as a single crystal). The latter can be associated with the elimination of SiC surface defects by surface restructuring during the sublimation growth. The in-plane crystallite size (L_a) is estimated from the ratio

A_D/A_G of the integrated intensities of the D band (A_D) and G band (A_G) through Raman spectroscopy [32]:

$$L_a = 2.4 \times 10^{-10} \lambda_{\text{laser,nm}}^4 \left(\frac{A_D}{A_G} \right)^{-1}$$

Micro-Raman spectroscopy imaging (μRSI, Fig. 3.20a–c) analysis has indicated that the graphene grown at 1800°C exhibits decoupled graphene layers, while at higher temperatures, the graphene stack consists mostly of graphitic layers. The transition from decoupled to Bernal-stacked graphene layers with the increase in growth temperature is attributed to a competition between growth mechanisms, i.e., between spontaneous growth and defect-assisted growth. At high temperature, the increase in the amount of C atoms leads to the formation of Bernal-stacked graphene layers around the extended surface defects such as polishing lines. Furthermore, the thickness dependence of the graphene dielectric function is deduced. The graphene grown at 1800°C shows graphene-like behavior, while the graphene grown at 2000°C has dielectric function closer to the one reported for graphite in accordance with the μRSI results. Independently, SE analysis and TEM revealed the presence of an interface layer between the graphene and the SiC substrate. We have found that this interface layer is amorphous and consists of a mix of C and Si, which are trapped during the sublimation process. The interface layer may have critical implications for the application of C-face graphene in future electronic devices.

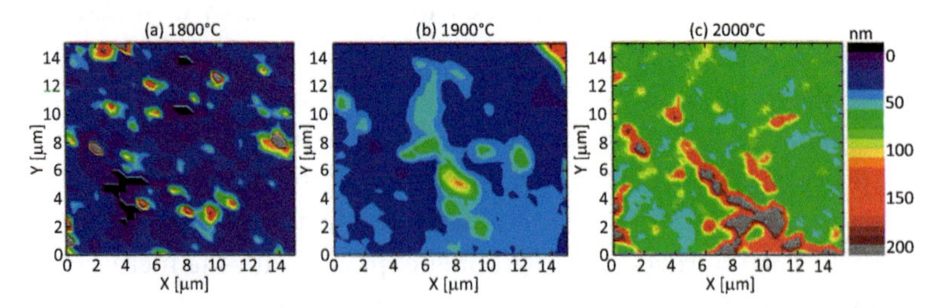

Figure 3.20 Three 15 μm × 15 μm color-coded maps of in-plane crystallite size for the graphene grown at (a) 1800°C, (b) 1900°C, and (c) 2000°C. The color bar represents the crystallite size in nm. The gray color in the bar represents a defect-free area. The step is 0.5 μm for the maps in (a) and (b) and 0.25 μm for the map in (c). Reprinted from Ref. [94], with permission from AIP Publishing LLC.

3.2.4 Growth on 3C-SiC and Its Comparison with other Polytypes

Both 4H-SiC and 6H-SiC have a hexagonal structure and, therefore, represent, in principle, an ideal template for graphene growth. In comparison with graphene growth on hexagonal SiC, limited attention has been given to graphene growth on 3C-SiC, as it has a cubic structure. The (111) surface of this crystal is naturally compatible with the sixfold symmetry of graphene. However, 3C-SiC grown on SiC is known to contain a lot of defects, which prevent this material of extensive electronic device developments.

In our study, we use 3C-SiC(111) grown on 6H-SiC(0001), which would eliminate thermal and lattice mismatch and opens a new possibility for applications. As an extension of graphene applications, the superior biocompatibility of 3C-SiC as compared to Si can also be used [95]. This may offer an attractive platform for the growth of graphene that could lead to a new generation of advanced biomedical devices. Additionally, graphene may be used as contact layers on solar cells made on 3C-SiC [96]. Surface restructuring after heating up SiC results in the formation of steps and terraces, which may have an impact on the doping uniformity (graphene conductance). The main effect behind the surface restructuring is the phenomenon of step bunching, which is distinctive of the different SiC polytypes, due to energy reasons.

We have reported a large-area monolayer EG on 3C-SiC(111) [17]. To better understand the material quality, 6H- and 4H-SiC substrates have been used as reference. Graphene formation has been analyzed as correlated to step bunching of SiC and taking into account the initial roughness of the substrate surface. The graphene samples grown at identical conditions on 4H-, 6H-, and 3C-SiC substrates were characterized by LEEM in order to evaluate thickness distribution. The bright area in all LEEM images (Fig. 3.21) represents an ML graphene, while the darker areas represent bilayer graphene. Figure 3.21a shows the LEEM image for graphene grown on a 4H-SiC substrate. Graphene on this sample consists of one and two MLs with one small area of three MLs (the black spot). Large and homogeneous ML graphene grown on 6H-SiC and 3C-SiC is shown in Fig. 3.21b,c. The areas of 1 ML coverage as extracted from the LEEM images are about 60%, 90%, and 98% for 4H, 6H, and

3C polytypes, respectively. During heating an SiC substrate above 1200°C, SiC surface undergoes microscopic restructuring by forming steps. This process, called step bunching, is different from surface reconstruction and refers to changing surface morphology. Step bunching, which is governed by energy minimization on different terraces, is a fundamental phenomenon in SiC.

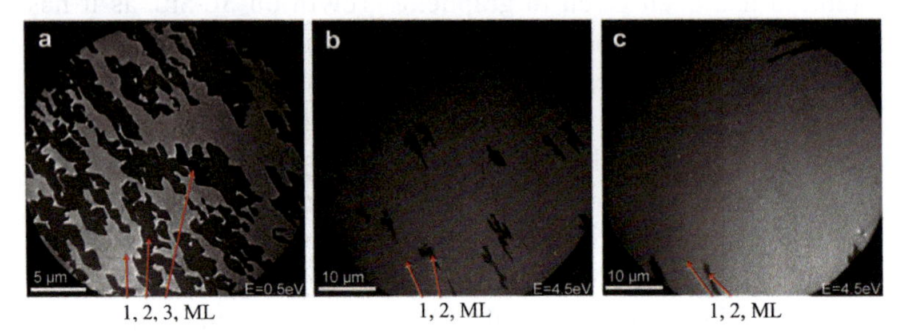

Figure 3.21 LEEM images of graphene on different SiC polytypes grown at identical conditions: (a) 4H-SiC with 60% coverage by ML, (b) 6H-SiC with 92% of 1 ML coverage, and (c) 3C-SiC with 98% 1 ML coverage. Reprinted from Ref. [17], with permission from Elsevier.

Several models have been proposed for the step bunching mechanism during SiC growth. Heine et al. considered that the energies of interaction for each SiC bilayer plane are different due to the unique stacking sequence of each polytype [97]. Kimoto et al. used Heine's calculation to discuss the formation of the unit cell height steps from a viewpoint of the surface equilibrium process in which the terrace with minimal stacking energy on a stepped surface will take over during SiC epitaxial growth [98]. In the case of SiC thermal decomposition, Si sublimation, we have a reverse scenario, which is now related to the disintegration of Si–C bilayers from the lattice stack. Since Si has the highest vapor pressure, Si leaves the surface, while C nominally rests and migrates on the surface. To study the surface restructuring during SiC substrate sublimation at 2000°C, we examined around 300 steps for each sample above using AFM. Figure 3.22(a–c) depicts histograms of the step height probability for the graphenized surfaces of 4H-, 6H-, and 3C-SiC substrates, respectively. Starting from a typical step height around 0.25 nm before heating, the steps grouped in four major heights related to the polytype structure. Some dispersion of the step heights above 1 nm was observed in all samples, but with a very low probability (not shown). Having a rather low step height distribution is one advantage of our results, since it has been reported that the resistance of EG on SiC increases linearly with step height on the substrate [99]. The corresponding histogram of the step height for 4H-SiC (Fig. 3.22a) indicates that two bilayer-height steps are the most probable and four bilayer-height steps show a significant probability. For the 6H-SiC sample (Fig. 3.22b), two and three bilayer-height steps dominate. On the 3C-SiC graphene sample, one Si–C bilayer height has the highest percentage (48%) of appearance, although some larger steps are present (Fig. 3.22c).

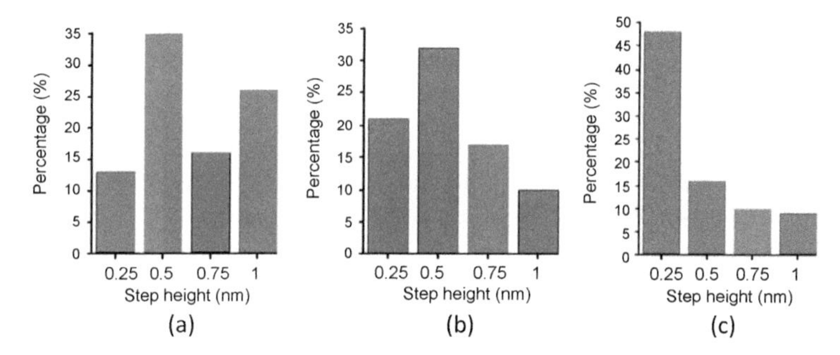

Figure 3.22 The respective histograms of the step height probability for the graphenized surfaces of (a) 4H-SiC, (b) 6H-SiC, and (c) 3C-SiC substrates. Reprinted from Ref. [17], with permission from Elsevier.

Based on the mentioned terrace energies (Section 3.2.2) for 4H-SiC, it will cost less energy to remove a 4H1 terrace. Thus, for the 4H1 terrace, the step-decomposition velocity will be faster (see Fig. 3.7a), and two different step heights of 0.5 nm and 1 nm are probable to be formed. A similar mechanism of energy minimization is expected in the 6H-SiC polytype (Fig. 3.7b). As a result, first the step 6H1 will catch step 6H2 and form two Si–C bilayers. Then step 6H3 will advance and merge with the two bilayer step. On 3C-SiC, all terraces have the same decomposition energy (Fig. 3.7c) and no energetically driven step bunching should be expected. In fact, the most probable step height observed (Fig. 3.22c) is 0.25 nm. On the image (Fig. 3.22c), one can see some additional step heights, which suggests that there are additional factors governing the surface restructuring. In 3C-SiC, a non-uniform sublimation can be induced by the presence of extended defects such as stacking faults characteristic of this material.

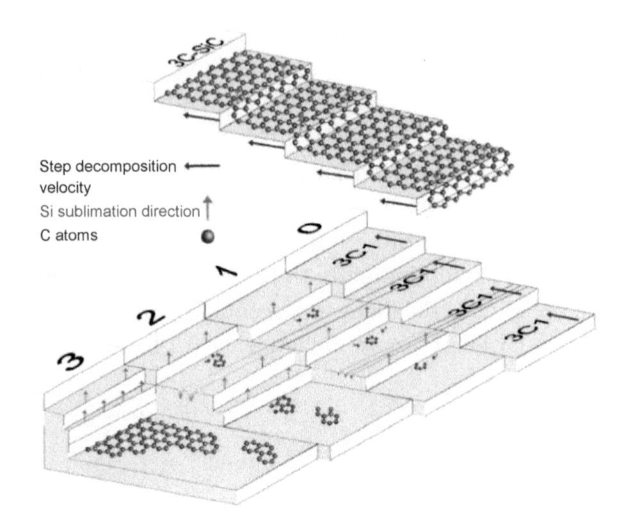

Figure 3.23 Graphene growth evolution (buffer layer formation not shown): top image on a defect-free 3C-SiC substrate. Bottom image on a 3C-SiC with defects (lines indicate stacking faults). Stage 0: steps with one Si–C bilayer height and the same terrace energy. Stage 1: sublimation of Si atoms from the edge of steps and ML graphene formation. Stage 2: V-shaped pits show erosion on stacking faults and subsequent step bunching. Stage 3: after step bunching of three Si–C bilayers, the source of C atoms increases and a larger graphene layer starts to form. Reprinted from Ref. [17], with permission from Elsevier.

A hypothetical model of graphene development on 3C-SiC (for 4H-SiC shown in Section 3.2.2) is depicted in Fig. 3.23, without accounting for the buffer layer formation. The decomposition rate of all 3C-SiC terraces is the same in a defect-free crystal, thus providing a uniform source of C on the surface (Fig. 3.23, top image), which results in a superior uniformity of the grown graphene layer (Fig. 3.21c). Note the frequency of formation of one bilayer step height over the surface for 3C sample in Fig. 3.22c. However, the presence of defects, e.g., stacking faults, with typical density of 5×10^3 cm^{-1} [100], may be a reason of step bunching on the 3C-SiC surface (Fig. 3.22c). Actually, step decomposition becomes faster at the defect sites (stages 1 and 2 in Fig. 3.23, bottom image) on the 3C-SiC surface, resulting in a non-uniform terrace removal and, therefore, defect-driven step bunching on the 3C-SiC surface (stage 3 in Fig. 3.23, bottom image). Our results also indicate that there exists a range of optimal terrace width that should be kept in order to maintain formation of one monolayer graphene and to avoid increase in carrier concentration [40].

Large-area μ-SE mapping (Fig. 3.24) and LEEM/μ-LEED investigations revealed critical correlations between surface reconstruction, graphene layer thickness, and electronic properties of EG grown on Si-face and C-face 3C-SiC [63]. Growth of single monolayer graphene is demonstrated on both Si- and C-faces, where large homogeneous domains with size up to ~2 × 2 mm^2 are achieved on the Si-face. On the C-face, the domains having homogeneous thickness are considerable smaller. The interface layer in this case also shows a distinctively different picture, with small uncorrelated nucleation sites that have high graphene coverage within the interface layer. We speculate that these sites may be associated with small pit defects on the C-face substrate that represent 6H-SiC spiral growth inclusions. Furthermore, the maps of the free carrier scattering time show that the carrier mobility in the homogeneous areas of 1 ML graphene is higher than the mobility in the thicker graphite islands. We find that the CP energy position associated with an exciton-enhanced van-Hove singularity in the density of states at ~4.5 eV blue shifts with decrease in the number of layers for both polarities. The analysis suggests that the interaction between graphene layer and the substrate is stronger for the Si-face material. These results are consistent with our LEED observations indicating the formation of a $(6\sqrt{3}\times 6\sqrt{3})R30^\circ$ buffer layer on the Si-face and absence of any specific reconstruction on the C-face samples.

We have studied defect formation in ~1 mm thick 3C-SiC layers grown on off-oriented 4H-SiC substrates via a lateral enlargement mechanism using different growth conditions [17]. In this method, a two-step growth process was developed, which provides a balance between the growth rate and the number of defects in the 3C-SiC layers. It proved that the two-step growth process combined with a geometrically controlled lateral enlargement mechanism allows the formation of a single 3C-SiC domain, which enlarges and completely covers the substrate surface. X-ray diffraction and low temperature photoluminescence measurements confirmed high crystalline quality of the grown 3C-SiC layers [101, 102]. Figure 3.25 illustrates the process conditions for 3C-SiC growth.

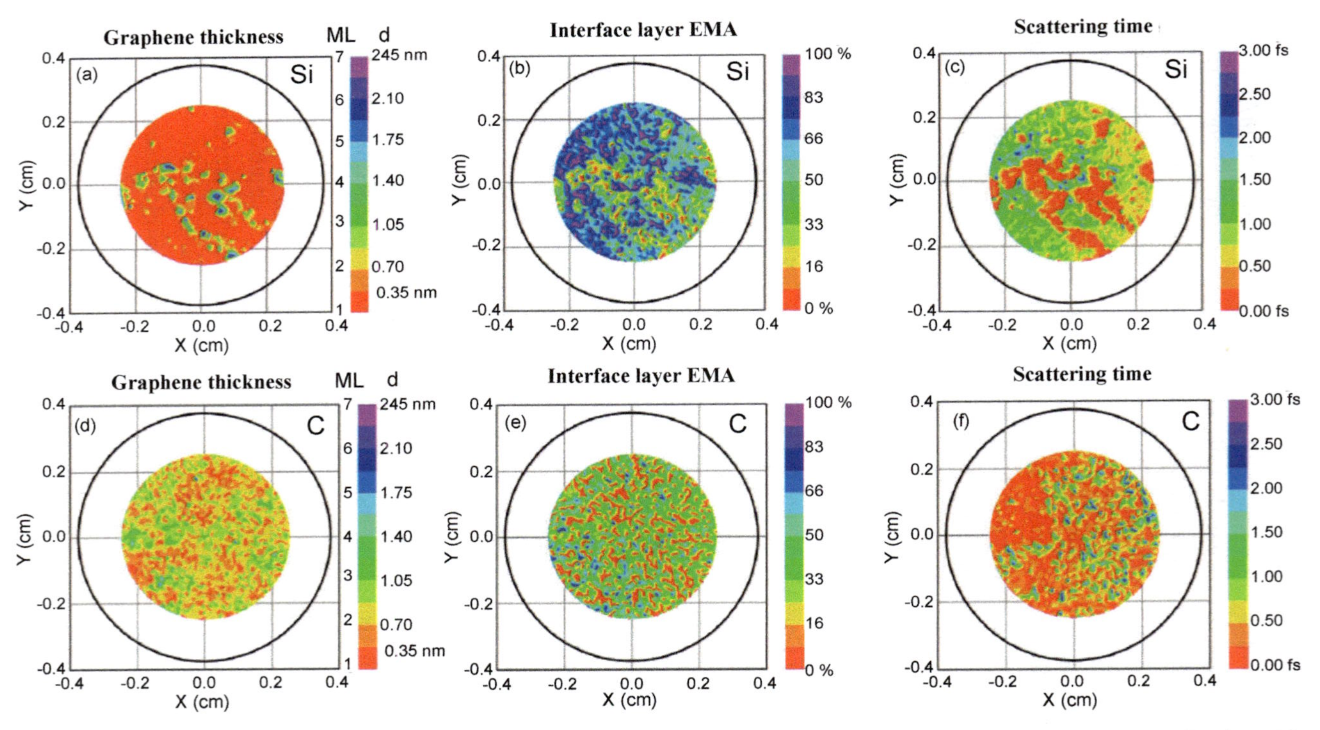

Figure 3.24 μ-SE maps of best-match model parameters for graphene layer thickness (a, d), interface-layer SiC percentage (b, e), and free-charge-carrier scattering time (c, f) for EG grown on Si-face (a, b, c) and C-face (d, e, f) of 3C-SiC. Reprinted from Ref. [63], with permission from AIP Publishing LLC.

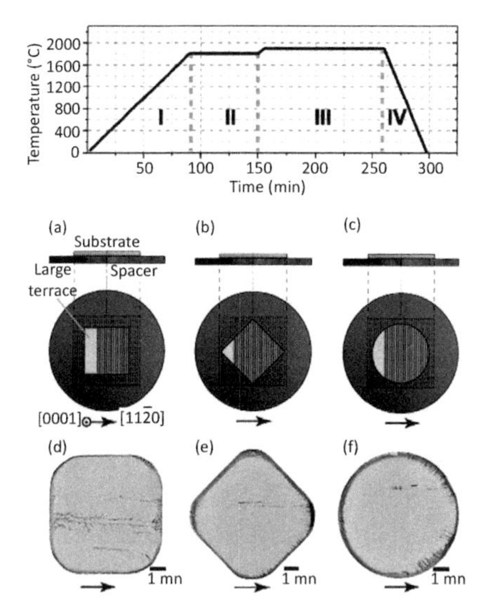

Figure 3.25 Sketch of two-step growth process of 3C-SiC layers (top). (a–c) Schematic illustration of spacer and substrate arrangements. Optical micrographs of the freestanding 3C-SiC grown using (d) original, (e) 45° rotated, and (f) circular spacer openings. Black arrows indicate the step-flow growth direction. Reprinted from from Refs. [101] and [102], with permission from American Chemical Society.

3.3 Graphene Growth on Etched SiC Substrates

3.3.1 Substrate Preparation by Etching

SiC wafers with mechanically polished surface are often damaged, rough, and showing a high density of scratches, as seen by AFM (Fig. 3.26a) [32]. Etching of SiC substrates prior to growth is an indispensable process in order to obtain scratch-free surfaces for subsequent epitaxial deposition. Hydrogen etching [103–107], thermal etching [108–111], and tetrafluorosilane (SiF_4) etching [112] are known to improve surface condition by removing several hundred nanometers of bulk material. In this section, we present some hydrogen etching studies.

In the CVD growth of SiC, hydrogen etching is a common procedure to prepare SiC substrate and provide a well-defined surface with atomically flat terraces for the growth of EG. A hydrogen etching study on the Si-face of SiC substrates showed that there is an optimal hydrogen flow (0.5 standard liters per minute, slpm) and sample temperature (1500°C) to achieve the best results [107]. Figure 3.27 shows AFM images of SiC samples etched in hydrogen under different process conditions. All samples were annealed for 15 min at an ambient hydrogen pressure of 1 atm. The temperature range was 1450–1550°C, and the hydrogen flow rate was varied between 0.2 and 3.0 slpm. Samples etched with a hydrogen flow ≤0.5 slpm and a temperature of 1450°C still show residual scratches, whereas etching with the same flow at 1550°C leads to droplet-like features at the SiC step edges. At 1550°C and a high flow of 3.0 slpm, step bunching and large defects were observed. The best hydrogen etching conditions were used to treat the SiC samples before graphitization, and graphene growth results on these samples show that the thickness distribution is Gaussian with a mean of 1.1 ML and a standard deviation of 0.17 ML. This shows that the growth of EG is very robust and reproducibility is due to the high controllability of the process parameters [107].

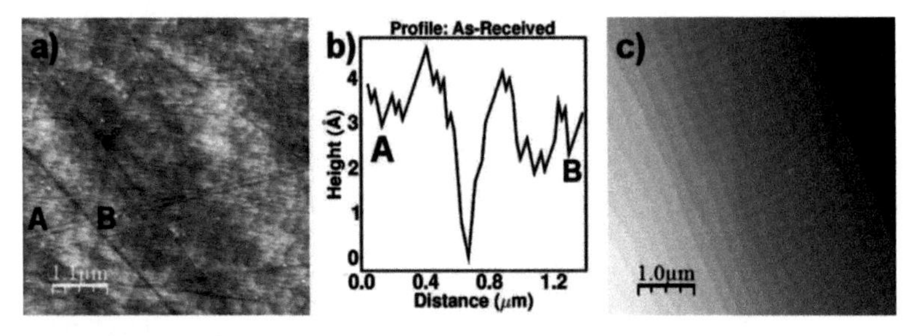

Figure 3.26 AFM image of an (a) as-received SiC(0001) substrate, (b) a line-profile from A to B, and (c) hydrogen-etched SiC. Reprinted from Ref. [106], with kind permission of Dr. Zachary Robinson, the College at Brockport, State University of New York, USA.

The C-face of SiC substrates was also etched in hydrogen at a temperature range of 1350–1550°C, for either 0 or 30 min [106]. All samples were etched at 200 mbar of hydrogen, with a flow rate of 5 slpm. The results showed that there is an optimal temperature (1450°C) and time (0 min). Instead, it was found that the 0 min etch at 1550°C and 30 min etch at 1450°C were resulting in a surface with a number of pits (Fig. 3.28a), while for 0 min etches at 1350°C and 1450°C, no pits were visible on the SiC surface (Fig. 3.28b). The AFM of this surface (Fig. 3.28d) reveals a much larger terrace width

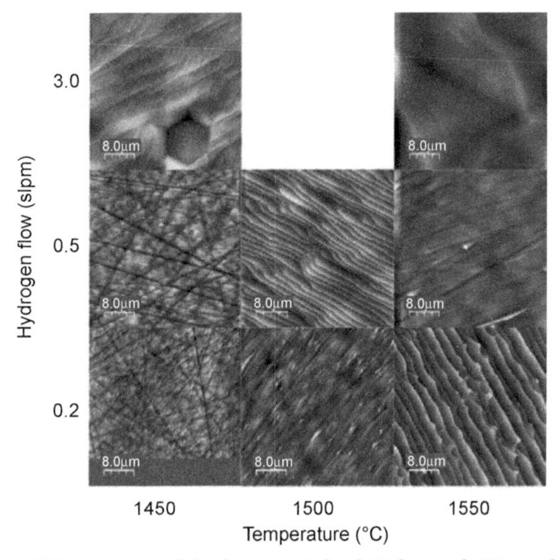

Figure 3.27 AFM images of hydrogen-etched Si-face of SiC surfaces etched with various hydrogen flow rates and temperatures. Reprinted from Ref. [107], with permission from WILEY-VCH Verlag GmbH & Co. KGaA, Weinheim.

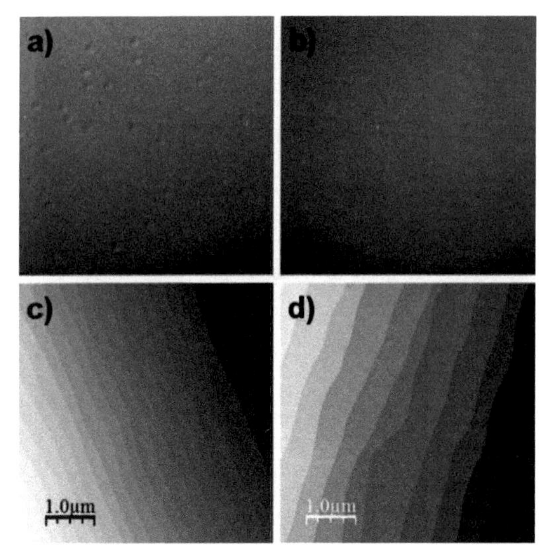

Figure 3.28 Nomarski and AFM images of etched SiC(0001). (a) Nomarski image of the SiC(0001) surface following a 30 min etch at 1450°C. The small features on the surface are pits. (b) Nomarski image of etched sample for 0 min. (c) AFM images of an area in (a) (TW ≈ 250 nm), (d) an area in (b) (TW ≈ 250 nm). Reprinted from Ref. [106], with kind permission of Dr. Zachary Robinson, the College at Brockport, State University of New York, USA. The optical images in (a) and (b) are both 150 µm × 150 µm.

of about ≈550 nm. The formation pits are due to the preferential etching of threading screw dislocations that intersect the surface. Additionally, on a flat area, AFM results show step morphology (Fig. 3.28c) with an average terrace width of 250 nm. Based on the results (larger terrace width, lack of pits), etching at 1450°C for 0 min was used for all the samples. The larger terrace widths are advantageous because they allow for fewer steps per electronic device, which results in improved performance.

3.3.2 Graphene Nanoribbons on SiC

Lack of bandgap in graphene and its behavior as a semi-metal hinder electronic applications of graphene, especially in switching/logic devices. There are different approaches to open the bandgap by breaking the graphene crystal symmetry. Theoretically, considering the confinement of electrons into one dimension [113], bandgap opening can be expected in graphene nanoribbons (GNRs). They can be produced as strips of nanometer width cut from graphene. In general, GNRs can be of two types depending on their edge termination. Figure 3.29 shows the structures of armchair and zigzag GNRs. The width of armchair GNRs is determined by the number of hexagonal carbon rings, or generally referred to as dimer lines (Na) across the ribbon. Similarly, the width of zigzag GNRs is dependent on the number of zigzag chains (Nz) across the ribbon [114].

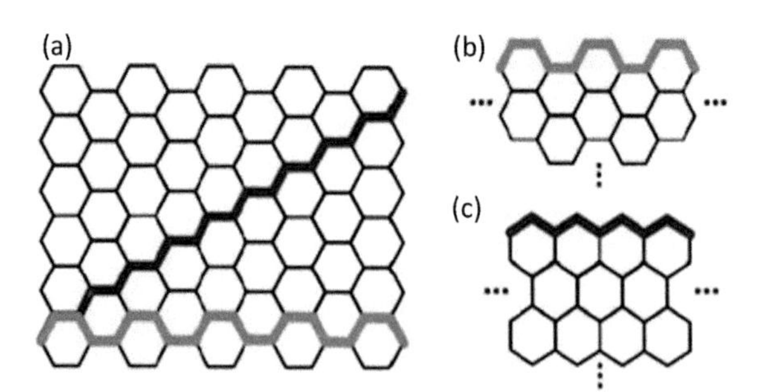

Figure 3.29 (a) Structure of graphene nanoribbon, (b) armchair, (c) zigzag [32].

Nakada et al., in 1966, studied theoretically the band structure of GNRs by using a tight-binding approach. They showed the importance of the edge type and the edge termination for the band structure [114]. The armchair nanoribbons are predicted to have a bandgap that is inversely proportional to the ribbon width. Mechanical cutting, etching of graphene, and other published methods to produce such nanoribbons are not appropriate because often the edges are destroyed and become disordered, which compromises the aimed properties [115–117]. A different approach was proposed by de Heer et al. in 2010 [118]. They explored the self-organized growth of graphene nanoribbons on a templated SiC substrate prepared using scalable photolithography and microelectronics processing. A precise control over the natural step bunching mechanism during SiC heating to elevated temperatures allowed the preparation of a crystal facet for self-organized graphene growth. The natural choice for this purpose may be (1–10n) facets. Controlled facets are achieved by the reactive ion etching of trenches of defined depths since it was found that it is the etch depth that ultimately defines the width of the GNRs prepared in the subsequent treatment. Etch depths of 20 nm were readily achieved, resulting in a ribbon width (~40 nm) sufficiently narrow to give a sizable bandgap at room temperature. The nanoribbon growth steps are depicted in Fig. 3.30.

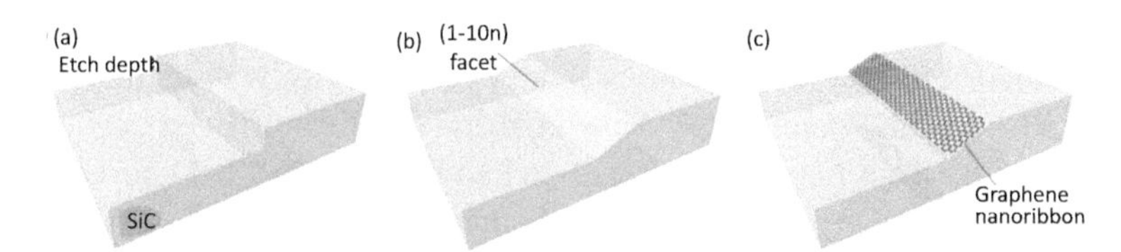

Figure 3.30 Process steps modifying the SiC substrate surface for graphene nanoribbon growth. (a) A nanometer-scale step is etched into SiC crystal by fluorine-based RIE. (b) The crystal is heated to 1200–1300°C (at low vacuum), inducing step flow and relaxation to the (1–10n) facet. (c) Upon further heating to ~1450°C, self-organized graphene nanoribbon forms on the facet. Reprinted from Ref. [118], with permission from John Wiley and Sons.

Direct nanoribbon growth avoids the need for damaging by post-growth processing. Raman spectroscopy, high resolution TEM, and EFM confirm that nanoribbons as narrow as 40 nm can be grown at specified positions on the substrate. Prototype graphene devices have been demonstrated to exhibit quantum confinement at 4 K and an on–off ratio of 10. Carrier mobility up to 2700 $cm^2/V\,s$ at room temperature was measured. They also demonstrate the scalability of this method by fabricating 10,000 top-gated graphene transistors on a 0.24 cm^2 SiC chip.

Other methods have been suggested, e.g., in Ref. [119]. The authors describe a technique for selective graphene growth and nanoribbon production onto 4H- and 6H-SiC. They demonstrated that graphene layers can be selectively grown onto Au- and Si-implanted SiC at pressures and temperatures of 1×10^{-6} Torr and 1200°C, respectively. Upon ion implantation, the graphitization temperature of SiC is lowered by at least 100°C, allowing selective growth of graphene layers on ion-implanted areas. The authors believe that by optimizing process conditions and starting with better-quality SiC, the growth of nanoribbons with reduced disorder can be achieved. However, they do not demonstrate any physical characteristics of such GNRs. In Ref. [120], a focused ion beam was used to pattern EG on SiC into an array of graphene nanoribbons as narrow as 15 nm by optimizing the Ga^+ ion beam current, acceleration voltage, dwell time, beam center to center distance, and ion dose. The ion dose required to completely etch away graphene on SiC was determined and compared with the Monte Carlo simulation result. In addition, a photodetector using an array of 300 of 20 nm graphene nanoribbons was fabricated and its photo-response was studied. The device's zero-bias photo-responsivity was estimated to be 7.32 mA/W. Recently, Wang et al. have grown nitrogen-seeded graphene from patterned stripes etched in the SiC(000-1) surface. The patterned growth produces a set of parallel nitrogen-seeded SiC stripes (400–500 nm wide). Graphene growth on these patterned surfaces leads to the formation of nitrogen-graphene ribbons with substantially improved structural and electronic properties [121].

3.4 Summary and Outlook

Graphene on SiC can be grown by thermal decomposition in conditions close to thermodynamic equilibrium. This provides good crystal quality, as related to point defects, on an area as large as the available SiC wafers, graphene deposition being now demonstrated up to 200 mm in diameter. Six-inch wafers are readily on the market and can be purchased from several suppliers.

When grown on the Si-terminated face of SiC, graphene lattice is commensurate with the buffer layer, having a negligible mismatch. This implies epitaxial conditions for graphene formation, which is another prerequisite of high structural quality. There are still two challenging issues for this type of graphene related to the influence of the substrate: (1) undesired doping induced by the substrate via the buffer layer and (2) step bunching phenomenon in SiC upon heating. What is more, these effects are interrelated to a certain extent. While the latter can be minimized by tuning the growth conditions or using cubic SiC, the former one has been tackled by applying hydrogen intercalation. On the other hand, graphene grown on the C-face is decoupled from the substrate and may have more successful applications once the material deposition is mastered.

The main problem arises from differences and variabilities in the growth mode; graphene is completed by domains with orientation disorder and hence non-uniform thickness results. Efforts are being made to overcome this difficulty, such as by applying various growth setup/configurations and operation parameters, such as growth conditions.

The graphene limitation of having no energy bandgap, which prevents graphene from application in, e.g., logic devices has been addressed by different means. One of them is the opening of a bandgap in graphene nanoribbons. Such structures would facilitate the realization of ballistic transport as well.

Recent progress in the growth of graphene with high carrier mobility, together with smart device designs, is expected to foster the application of EG in electronic devices. Progress in this direction relies on the advance of well-controlled growth technologies that enable thickness uniformity of the graphene layer, as well as reproducible mobility and carrier concentration at wafer-scale SiC (>150 mm in diameter). Continuous exploration of its unique physical phenomena will increase the potential of graphene on SiC technology for microelectronics far beyond quantum metrology, which is the most established application area of this graphene to date.

Acknowledgments

The research leading to these results has received funding from the European Union Seventh Framework Program under grant agreement n°604391 Graphene Flagship. Authors would like to acknowledge the Swedish Research Council for financial support via contracts VR 621-2014-5805 and the LiU Linnaeus Grant.

References

1. Boehm, H. P. (1962). Dunnste kohlenstoff-folien, *Z. Naturf*, **17**, pp. 150–153.
2. van Bommel, A. J. (1975). LEED and Auger electron observation, *Surf. Sci.*, **48**, pp. 463–472.
3. Lu, X. K. (1999). Tailoring graphite with the goal of achieving single sheets, *Nanotechnology*, **10**, pp. 269–272.
4. Novoselov, K. S. (2004). Electric field effect in atomically thin carbon films, *Science*, **306**, pp. 666–669.
5. Partoens, B. (2006). From graphene to graphite: Electronic structure around the K point, *Phys. Rev. B*, **74**, pp. 075404.
6. Novoselov, K. S. (2005). Two-dimensional gas of massless Dirac fermions in graphene, *Nature*, **438**, pp. 197–200.
7. Castro Neto, A. H. (2009). The electronic properties of graphene, *Rev. Mod. Phys.*, **81**, pp. 109–162.
8. Lin, Y. M. (2010). 100-GHz transistors from wafer-scale epitaxial graphene, *Science*, **327**, pp. 662.
9. Pearce, R. (2011). Epitaxially grown graphene based gas sensors for ultra sensitive NO_2 detection, *Sens. Actuators B*, **155**, pp. 451–455.
10. Tzalenchuk, A. (2010). Towards a quantum resistance standard based on epitaxial graphene, *Nat. Nanotechnol.*, **5**, pp. 186–189.
11. Lang, B. (1975). A LEED study of the deposition of carbon on platinum crystal surfaces, *Surf. Sci.*, **53**, pp. 317–329.
12. Wu, Y. H. (2010). Two-dimensional carbon nanostructures: Fundamental properties, synthesis, characterization, and potential applications, *J. Appl. Phys*, **108**, pp. 071301.
13. Li, X. (2009). Evolution of graphene growth on Ni and Cu by carbon isotope labeling, *Nano Lett.*, **9**, pp. 4268–4272.
14. Sutter, P. W. (2008). Epitaxial graphene on ruthenium, *Nat. Mater.*, **7**, pp. 406–411.
15. Somani, P. R. (2006). Planer nano-graphenes from camphor by CVD, *Chem. Phys. Lett.*, **430**, pp. 56–59.

16. Yakimova, R. (2010). Analysis of the formation conditions for large area epitaxial graphene on SiC substrates, *Mater. Sci. Forum,* **645–648**, pp. 565–568.

17. Yazdi, G. R. (2013). Growth of large-area monolayer graphene on 3C-SiC and a comparison with other SiC polytypes, *Carbon,* **57**, pp. 477–484.

18. Berger, C. (2006). Electronic confinement and coherence in patterned epitaxial graphene, *Science,* **312**, pp. 1191–1196.

19. Ohta, T. (2006). Controlling the electronic structure of bilayer graphene, *Science,* **313**, pp. 951–954.

20. Riedl, C. (2007). Structural properties of the graphene-SiC(0001) interface as a key for the preparation of homogeneous large-terrace graphene surfaces, *Phys. Rev. B,* **76**, pp. 245406.

21. Emtsev, K. V. (2009). Towards wafer-size graphene layers by atmospheric pressure graphitization of silicon carbide, *Nat. Mater.,* **8**, pp. 203–207.

22. Tairov, Y. M. (1978). Investigation of growth processes of ingots of silicon carbide single crystals, *J. Cryst. Growth*, **43**, pp. 209–212.

23. Tairov, Y. M. (1981). General principles of growing large-size single crystals of various silicon carbide polytypes, *J. Cryst. Growth*, **52**, pp. 146–150.

24. Lilov, S. (1993). Study of the equilibrium processes in the gas phase during silicon carbide sublimation, *Mater. Sci. Eng. B*, **21**, pp. 65–69.

25. Jernigan, G. G. (2009). Comparison of epitaxial graphene on Si-face and C-face 4H-SiC formed by ultrahigh vacuum and RF furnace production, *Nano Lett.*, **9**, pp. 2605–2609.

26. Berger, C. (2004). Ultrathin epitaxial graphite: 2D electron gas properties and a route toward graphene-based nanoelectronics, *J. Phys. Chem. B*, **108**, pp. 19912–19916.

27. Guo, Z. (2013). Record maximum oscillation frequency in C-face epitaxial graphene transistors, *Nano Lett.*, **13**, pp. 942–947.

28. Baringhaus, J. (2014). Exceptional ballistic transport in epitaxial graphene nanoribbons, *Nature*, **506**, pp. 349–354.

29. Hicks, J. (2013). A wide-bandgap metal-semiconductor-metal nanostructure made entirely from graphene, *Nat. Phys.*, **9**, pp. 49–54.

30. Drowart, J. (1958). Thermodynamic study of SiC utilizing a mass spectrometer, *J. Chem. Phys.*, **29**, pp. 1015–1021.

31. Forbeaux, I. (1998). Heteroepitaxial graphite on 6HSiC(0001): Interface formation through conduction-band electronic structure, *J. Phys. Rev. B*, **58**, pp. 16396–16406.
32. Yazdi, G. R. (2013). Epitaxial graphene on SiC: A review of growth and characterization, *Crystals*, **6**, pp. 53.
33. Berzelius, J. J. (1824). Untersuchungen über die Flusspath Faure und deren merkwürdige Verbindungen, *Ann. Phys., Lpz.*, **1**, pp. 169–230. (In German)
34. Acheson, A. G. (1892). England Patent, 17911.
35. Acheson, A. G. (1893). On carborundum, *Chem. News*, **68**, pp. 179.
36. Round, H. J. (1907). A note on carborundum, *Electr. World*, **19**, pp. 309–310.
37. Lely, J. A. (1955). Darstellung von Einkristallen von Silicium carbid und Beherrschung von Art und Menge der eingebautem Verunreingungen, *Ber. Deut. Keram. Ges*, **32**, pp. 229–250.
38. Ramsdell, L. S. (1947). Studies on silicon carbide, *Am. Min*, **32**, pp. 64–82.
39. Verma, A. R. (1966). *Polymorphism and Polytypism in Crystals*, John Wiley & Sons Inc., New York.
40. Yakimova, R. (2014). Morphological and electronic properties of epitaxial graphene on SiC, *Phys. B*, **439**, pp. 54–59.
41. Johansson, L. I. (2014). Properties of epitaxial graphene grown on C-face SiC compared to Si-face, *J. Mater. Res.*, **29**, pp. 426–438.
42. Norimatsu, W. (2014). Structural features of epitaxial graphene on SiC{0001} surfaces, *J. Phys. D Appl. Phys.*, **47**, pp. 094017.
43. Chien, F. R. (1994). Terrace growth and polytype development in epitaxial SiC on -SiC (6H and 15R) substrates, *J. Mater. Res.*, **9**, pp. 940.
44. Ohta, T. (2010). Role of carbon surface diffusion on the growth of epitaxial graphene on SiC, *Phys. Rev. B*, **81**, pp. 121411.
45. Borovikov, V. (2009). Step bunching of vicinal 6H-SiC{0001} surfaces, *Phys. Rev. B*, **79**, pp. 245413.
46. Bolen, M. (2009). Graphene formation mechanisms on 4H-SiC(0001), *Phys. Rev. B*, **80**, pp. 115433.
47. Ming, F. (2011). Model for the epitaxial growth of graphene on 6H-SiC(0001), *Phys. Rev. B*, **84**, pp. 115459.
48. Virojanadara, C. (2008). Homogeneous large-area graphene layer growth on 6H-SiC(0001), *Phys. Rev. B*, **78**, pp. 245403.

49. Riedl, C. (2010). Structural and electronic properties of epitaxial graphene on SiC(0001): A review of growth, characterization, transfer doping and hydrogen intercalation, *J. Phys. D*, **43**, pp. 374009.
50. de Heer, W. A. Berger, C., Ruan, M., Sprinkle, M., Li, X., Hu, Y., Zhang, B., Hankinson, J., and Conrad, E. (2011). Large area and structured epitaxial graphene produced by confinement controlled sublimation of silicon carbide, *Proc. Natl. Acad. Sci. U.S.A.*, **108**, pp. 16900–16905.
51. Srivastava, N. (2012). Interface structure of graphene on SiC(000-1), *Phys. Rev. B*, **85**, pp. 041404.
52. Badami, D. V. (1962). Graphitization of silicon carbide, *Nature*, **193**, pp. 569–570.
53. Badami, D. V. (1965). X-Ray studies of graphite formed by decomposing silicon carbide, *Carbon*, **3**, pp. 53–57.
54. Charrier, A. (2002). Solid-state decomposition of silicon carbide for growing ultra-thin heteroepitaxial graphite films, *J. Appl. Phys.*, **92**, pp. 2479–2481.
55. Hibino, H. (2008). Microscopic thickness determination of thin graphite films formed on SiC from quantized oscillation in reflectivity of low-energy electrons, *Phys. Rev. B*, **77**, pp. 075413.
56. Ohta, T. (2008). Morphology of graphene thin film growth on SiC(0001), *New J. Phys.*, **10**, pp. 023034.
57. Hannon, J. B. (2008). Pit formation during graphene synthesis on SiC(0001): In situ electron microscopy, *Phys. Rev. B*, **77**, pp. 241404(R).
58. Hupalo, M. (2009). Growth mechanism for epitaxial graphene on vicinal 6H-SiC(0001) surfaces: A scanning tunneling microscopy study, *Phys. Rev. B*, **80**, pp. 041401(R).
59. Tromp, R. M. (2009). Thermodynamics and kinetics of graphene growth on SiC(0001), *Phys. Rev. Lett.*, **102**, pp. 106104.
60. Rutter, G. M. (2007). Imaging the interface of epitaxial graphene with silicon carbide via scanning tunneling microscopy, *Phys. Rev. B*, **76**, pp. 235416.
61. Nie, S. (2009). Tunneling spectroscopy of graphene and related reconstructions on SiC(0001), *J. Vac. Sci. Technol. A*, **27**, pp. 1052.
62. Emtsev, K. V. (2008). Interaction, growth, and ordering of epitaxial graphene on SiC{0001} surfaces: A comparative photoelectron spectroscopy study, *Phys. Rev. B*, **77**, pp. 155303.
63. Darakchieva, V. (2013). Large-area microfocal spectroscopic ellipsometry mapping of thickness and electronic properties of

epitaxial graphene on Si- and C-face of 3C-SiC(111), *Appl. Phys. Lett.*, **102**, pp. 213116.

64. Riedl, C. (2009). Quasi freestanding epitaxial graphene on SiC obtained by hydrogen intercalation, *Phys. Rev. Lett.*, **103**, pp. 246804.
65. Mattausch, A. (2007). Ab initio study of graphene on SiC, *Phys. Rev. Lett.*, **99**, pp. 076802.
66. Hannon, J. B. (2011). Direct measurement of the growth mode of graphene on SiC(0001) and SiC(0001), *Phys. Rev. Lett.*, **107**, pp. 166101.
67. de Lima, L. H. (2013). Atomic surface structure of graphene and its buffer layer on SiC(0001): A chemical-specific photoelectron diffraction approach, *Phys. Rev. B*, **87**, pp. 081403.
68. Borysiuk, J. (2009). Transmission electron microscopy and scanning tunneling microscopy investigations of graphene on 4H-SiC(0001), *J. Appl. Phys.*, **105**, pp. 023503.
69. Norimatsu, W. (2009). Transitional structures of the interface between graphene and 6H-SiC(0001), *Chem. Phys. Lett.*, **468**, pp. 52–56.
70. Norimatsu, W. (2010). Selective formation of ABC-stacked graphene layers on SiC(0001), *Phys. Rev. B*, **81**, pp. 161410(R).
71. Sonde, S. (2010). Uniformity of epitaxial graphene on on-axis and off-axis SiC probed by Raman spectroscopy and nanoscale current mapping, *Mater. Sci. Forum*, **645–648**, pp. 607–610.
72. Sun, G. F. (2009). Atomic-scale imaging and manipulation of ridges on epitaxial graphene on 6H-SiC(0001), *Nanotechnology*, **20**, pp. 355701.
73. Vecchio, C. (2011). Nanoscale structural characterization of epitaxial graphene grown on off-axis 4H-SiC(0001), *Nanoscale Res. Lett.*, **6**, pp. 269.
74. Berger, C. (2008). Dirac particles in epitaxial graphene films grown on SiC, *Adv. Solid State Phys.*, **47**, pp. 145–157.
75. Virojanadara, C. (2010). Buffer layer free large area bi-layer graphene on SiC(0001), *Surf. Sci.*, **604**, pp. L4–L7.
76. Xia, C. (2013). Detailed studies of Na intercalation on furnace-grown graphene on 6H-SiC(0001), *Surf. Sci.*, **613**, pp. 88–94.
77. Oliveira, M. H. (2013). Formation of high-quality quasi-freestanding bilayer graphene on SiC(0001) by oxygen intercalation upon annealing in air, *Carbon*, **52**, pp. 83–89.
78. Virojanadara, C. (2010). Epitaxial graphene on 6H-SiC and Li intercalation, *Phys. Rev. B*, **82**, pp. 205402.

79. Xia, C. (2012). Si intercalation/deintercalation of graphene on 6H-SiC(0001), *Phys. Rev. B*, **85**, pp. 045418.

80. Gierz, I. (2010). Electronic decoupling of an epitaxial graphene monolayer by gold intercalation, *Phys. Rev. B*, **81**, pp. 235408.

81. Walter, A. L. (2011). Highly p-doped epitaxial graphene obtained by fluorine intercalation, *Appl. Phys. Lett.*, **98**, pp. 184102.

82. Emtsev, K. V. (2011). Ambipolar doping in quasifree epitaxial graphene on SiC(0001) controlled by Ge intercalation, *Phys. Rev. B*, **84**, pp. 125423.

83. Tanabe, S. (2012). Quantum Hall effect and carrier scattering in quasi-freestanding monolayer graphene, *Appl. Phys. Express*, **5**, pp. 125101.

84. Giannazzo, F. (2012). Electronic transport at monolayer–bilayer junctions in epitaxial graphene on SiC, *Phys. Rev. B*, **86**, pp. 235422.

85. Mikoushkin, V. M. (2015). Size confinement effect in graphene grown on 6H-SiC (0001) substrate, *Carbon*, **86**, pp. 139–145.

86. Borysiuk, J. (2010). Transmission electron microscopy investigations of epitaxial graphene on C-terminated 4H-SiC, *J. Appl. Phys.*, **108**, pp. 013518.

87. Hass, J. (2008). Why multilayer graphene on 4H-SiC(0001) behaves like a single sheet of graphene, *Phys. Rev. Lett.*, **100**, pp. 125504.

88. Biedermann, L. B. (2009). Insights into few-layer epitaxial graphene growth on 4H-SiC(0001) substrates from STM studies, *Phys. Rev. B*, **79**, pp. 125411.

89. Lin, Y. (2010). Multicarrier transport in epitaxial multilayer graphene, *Appl. Phys. Lett.*, **97**, pp. 112107.

90. Siegel, D. A. (2010). Quasi-freestanding multilayer graphene films on the carbon face of SiC, *Phys. Rev. B*, **81**, pp. 241417.

91. Johansson, L. J. (2011). Stacking of adjacent graphene layers grown on C-face SiC, *Phys. Rev. B*, **84**, pp. 125405.

92. Johansson, L. J. (2013). Is the registry between adjacent graphene layers grown on C-face SiC different compared to that on Si-face SiC, *Crystals*, **3**, pp. 1–13.

93. Cristina, E. (2014). Exploring graphene formation on the C-terminated face of SiC by structural, chemical and electrical methods, *Carbon*, **69**, pp. 221–229.

94. Bouhafs, C. (2015). Structural properties and dielectric function of graphene grown by high-temperature sublimation on 4H-SiC(0001), *J. Appl. Phys.*, **117**, pp. 085701.

95. Saddow, S. E. (2011). Single-crystal silicon carbide: A biocompatible and hemocompatible semiconductor for advanced biomedical applications, *Mater. Sci. Forum,* **679–680**, pp. 824–830.
96. Beaucarne, G. (2002). The impurity photovoltaic effect in wide bandgap semiconductors: An opportunity for very-high-efficiency solar cells, *Prog. Photovolt. Res. Appl.,* **10**, pp. 345–353.
97. Heine, V. (1991). The preference of silicon carbide for growth in the metastable cubic form, *J. Am. Ceram. Soc.,* **74**, pp. 2630–2633.
98. Kimoto, T. (1995). Step bunching in chemical vapor deposition of 6H- and 4H-SiC on vicinal SiC(0001) faces, *Appl. Phys. Lett.,* **66**, pp. 3645–3647.
99. Low, T. (2012). Formation and scattering in graphene over substrate steps, *Phys. Rev. Lett.,* **108**, pp. 096601.
100. Vasiliauskas, R. (2011). Effect of initial substrate conditions on growth of cubic silicon carbide, *J. Cryst. Growth,* **324**, pp. 7–14.
101. Jokubavicius, V. (2014). Lateral enlargement growth mechanism of 3C-SiC on off-oriented 4H-SiC substrates, *Cryst. Growth Des.,* **14**, pp. 6514–6520.
102. Jokubavicius, V. (2015). Single domain 3C-SiC growth on off-oriented 4H-SiC substrates, *Cryst. Growth Des.,* **15**, pp. 2940–2947.
103. Chu, T. L. (1965). Chemical etching of silicon carbide with hydrogen, *J. Electrochem. Soc.,* **112**, pp. 955–956.
104. Owman, F. (1996). Removal of polishing-induced damage from 6H-SiC(0001) substrates by hydrogen etching, *J. Cryst. Growth,* **167**, pp. 391–395.
105. Robinson, Z. R. (2015). Challenges to graphene growth on SiC(0001): Substrate effects, hydrogen etching and growth ambient, *Carbon,* **81**, pp. 73–82.
106. Robinson, Z. R., Jernigan, G. G., Bussmann, K. M., Nyakiti, L. O., Garces, N. Y., Nath, A., Wheeler, V. D., Myers-Ward, R. L., Gaskill, D. K., and Eddy, C. R. (2015). Graphene growth on SiC(000): Optimization of surface preparation and growth conditions, *Proc. Carbon Nanotubes, Graphene, and Emerging 2D Materials for Electronic and Photonic Devices* VIII, 95520Y (in USA).
107. Ostler, M. (2010). Automated preparation of high-quality epitaxial graphene on 6H-SiC(0001), *Phys. Status Solidi B,* **247**, pp. 2924–2926.
108. Swiderski, I. (1972). Thermal etching of α-SiC crystals in argon, *J. Cryst. Growth,* **16**, pp. 1–9.

109. Nishiguchi, T. (2003). Thermal etching of 6H-SiC substrate surface, *Jpn. J. Appl. Phys.*, **42**, pp. 1533–1537.

110. van der Berga, N. G. (2012). Thermal etching of SiC, *Appl. Surf. Sci.*, **258**, pp. 5561–5566.

111. Lebedev, S. P. (2011). Formation of periodic steps on 6H-SiC(0001) surface by annealing in a high vacuum, *Mater. Sci. Forum*, **679**, pp. 437–440.

112. Rana, T. (2013). Vapor phase surface preparation (etching) of 4H-SiC substrates using tetrafluorosilane (SiF_4) in a hydrogen ambient for SiC epitaxy, *J. Cryst. Growth*, **380**, pp. 61–67.

113. Nakada, K. (1996). Edge state in graphene ribbons: Nanometer size effect and edge shape dependence, *Phys. Rev. B*, **54**, pp. 17954–17961.

114. Kan, E. (2011). Graphene nanoribbons: Geometric, electronic, and magnetic properties, *Intech Open J.*, pp. 331–348.

115. Han, M. Y. (2010). Electron transport in disordered graphene nanoribbons, *Phys. Rev. Lett.*, **104**, pp. 056801.

116. Tapaszto, L. (2008). Tailoring the atomic structure of graphene nanoribbons by scanning tunnelling microscope lithography, *Nat. Nanotech.*, **3**, pp. 397–401.

117. Jiao, L. (2009). Narrow graphene nanoribbons from carbon nanotubes, *Nature*, **458**, pp. 877–880.

118. Sprinkler, M. (2010). Scalable templated growth of graphene nanoribbons on SiC, *Nat. Nanotechnol.*, **5**, pp. 727–731.

119. Tongay, S. (2012). Drawing graphene nanoribbons on SiC by ion implantation, *Appl. Phys. Lett.*, **100**, pp. 073501.

120. Zhang, Y. (2014). A large-area 15 nm graphene nanoribbon array patterned by a focused ion beam, *Nanotechnology*, **25**, pp. 135301.

121. Wang, F. (2015). Pattern induced ordering of semiconducting graphene ribbons grown from nitrogen-seeded SiC, *Carbon*, **82**, pp. 360–367.

Chapter 4

Nanoscale Electrical and Structural Properties of Epitaxial Graphene Interface with SiC(0001)

Filippo Giannazzo,[a] Ioannis Deretzis,[a] Antonino La Magna,[a] Giuseppe Nicotra,[a] Corrado Spinella,[a] Fabrizio Roccaforte,[a] and Rositza Yakimova[b]

[a]*Istituto per la Microelettronica e Microsistemi Strada VII n.5 Zona Industriale, 95100 Catania, Italy*

[b]*Department of Physics, Chemistry and Biology, Linköping University, SE-58183 Linköping, Sweden*

filippo.giannazzo@imm.cnr.it

This chapter reports recent experimental investigations on the electrical and structural properties of the epitaxial graphene interface with the SiC(0001). Nanoscale resolution electrical characterization techniques based on scanning probe microscopy and atomic resolution structural analyses based on aberration-corrected scanning transmission electron microscopy have been jointly applied to elucidate the correlation between the interface structure and local transport properties (resistance, electron mean free path) of epitaxial graphene. A particular focus will be provided on the impact of nanosteps/facets of vicinal SiC surfaces

Epitaxial Graphene on Silicon Carbide: Modeling, Devices, and Applications
Edited by Gemma Rius and Philippe Godignon

ISBN 978-981-4774-20-8 (Hardcover), 978-1-315-18614-6 (eBook)
www.panstanford.com

and of lateral inhomogeneities in the number of graphene layers on the local resistance of epitaxial graphene. The results of these nanoscale investigations have been also compared with the results of the electrical characterization of macroscopic device structures fabricated with epitaxial graphene, clarifying the origin of some commonly observed effects, such as the anisotropic current transport with respect to steps orientation. Local transport measurements (in the diffusive regime) indicated a significantly higher electron mean free path for graphene residing on (11-2n) SiC facets than on the (0001) basal plane, consistently with recent reports on exceptional ballistic transport in epitaxial graphene nanoribbons.

4.1 Introduction

The controlled graphitization of hexagonal silicon carbide (SiC) during high-temperature thermal processes [1–3] represents one of the main approaches to obtain high-quality graphene on wafer scale for high-frequency electronics [4, 5], high-precision metrology [6], and environmental sensing [7]. The main advantage of this growth method is providing graphene directly on a semiconductor or semi-insulating substrate, i.e., ready for fabrication of electronic devices. To date, this approach has been widely investigated on the hexagonal (4H and 6H) polytypes of SiC [1–3]. Growth on the cubic (3C) polytype has also been explored [8], due to the possibility of integrating graphene with 3C-SiC layers on silicon. However, the high defectivity of 3C-SiC strongly affects the quality of graphene grown on its surface.

The growth mechanisms and, ultimately, the final structural and electronic properties (defectivity, thickness homogeneity, doping, carrier mobility) of graphene critically depend on the surface morphology and atomic termination of the substrate. This aspect is particularly relevant in the case of hexagonal SiC substrates, where graphene films with very different properties have been obtained on the different crystal orientations, i.e., the Si-face (0001) [9], the C-face (000-1) [9], and the low-index nonpolar faces (11–20) and (1–100) [10, 11]. Furthermore, the nanometric steps or facets of vicinal SiC surfaces have been shown to play an important role on the electrical properties of epitaxial graphene [12].

This chapter investigates the morphological, structural, and electrical properties of epitaxial graphene on the stepped/faceted surface of vicinal SiC(0001) wafers. In particular, atomic force microscopy (AFM) was used to study the morphology of epitaxial graphene grown on SiC(0001) samples with different miscut angles, while scanning transmission electron microscopy (STEM) and electron energy loss spectroscopy (EELS) provided atomic resolution structural information on graphene interface with the SiC basal plane and facets. Nanoscale electrical characterization by scanning probe microscopy (SPM) techniques has been extensively applied to investigate the effect of substrate steps/facets, as well as the effect of the lateral inhomogeneity in the number of graphene layers (e.g., monolayer/bilayer junctions) on the local resistance of epitaxial graphene films.

These nanoscale electrical analyses have been correlated with the results of the macroscopic electrical characterization of epitaxial graphene device structures (TLM, MOSFETs), providing the explanation of typically observed effects, such as the anisotropic current transport with respect to SiC steps orientation.

4.2 Epitaxial Graphene on Different SiC Orientations

Graphene growth by thermal decomposition of hexagonal SiC relies on the interplay of several mechanisms, i.e., the preferential Si sublimation from the topmost SiC layers, leaving an excess of carbon, the diffusion of these C atoms on the SiC surface, and their reorganization in the two-dimensional (2D) hexagonal graphene lattice. These mechanisms depend both on the annealing conditions (sample temperature, gas partial pressures in the chamber) and on the SiC surface termination.

In Fig. 4.1, a scheme of a 4H-SiC crystal is reported, showing the low-index (0001), (000-1), (11-20), and (1-100) crystallographic planes. In particular, (0001) and (000-1) surfaces are single-element terminated planes, referred to as Si-face or C-face, respectively. The displacement between Si and C atoms along these directions results in an electric polarization of the SiC crystal. As a consequence, (0001) and (000-1) surfaces are also commonly referred to as polar faces.

On the contrary, Si and C atoms are almost lying on the same plane in the case of (1–100) and (11–20) surfaces, which are typically called nonpolar faces. The different configurations of these low-index planes result in different surface reconstructions during thermal decomposition and, ultimately, in the formation of graphene films with very different structural, morphological, and electrical quality.

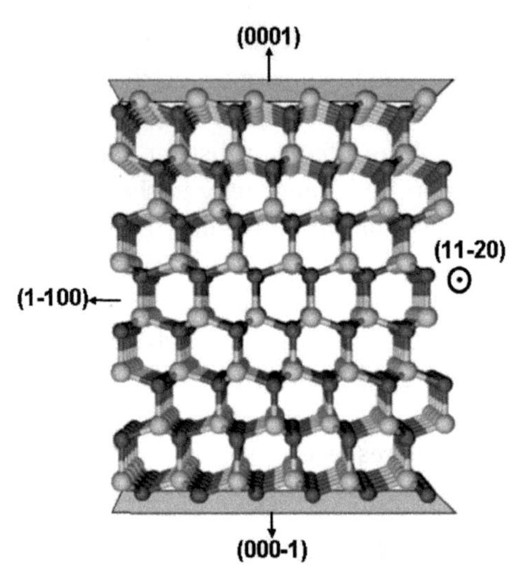

Figure 4.1 Schematic representation of a 4H-SiC crystal, showing the low-index (0001), (000-1), (11–20), and (1–100) crystallographic planes. Reprinted with permission from Ref. [11], Copyright 2013, American Physical Society.

To date, most of the studies on graphene growth on SiC have been performed on the (0001) and (000-1) polar faces, due to the availability of high-quality SiC substrates with these orientations. In particular, the growth on the (000-1) face is typically difficult to control and yields a laterally inhomogeneous distribution of graphene multilayers, in which the graphene layers are reciprocally misoriented and lacks of epitaxial relation with the substrate. On the contrary, a very precise epitaxial alignment with the substrate and an excellent uniformity in the number of layers (from one to a few layers with Bernal stacking sequence) have been demonstrated in the case of the (0001) face [2, 3]. For this reason, epitaxial graphene growth on 6H- or 4H-SiC (0001) has been the object of several investigations in the last years, which provided a thorough description of the SiC surface reconstruction sequence leading to graphene formation. In particular, a carbon buffer layer (BL) with $(6\sqrt{3} \times 6\sqrt{3})R30°$ reconstruction represents the precursor of graphene formation on the (0001) face of SiC [13]. In spite of its hexagonal lattice structure, the BL exhibits very different electronic properties with respect to graphene. In fact, part of the Si atoms of the (0001) face are covalently bonded to the BL, making it partially sp^3 hybridized, while a large density of dangling bonds is associated to the unsaturated Si atoms' bonds. Figure 4.2a depicts a schematic representation of the interface between the BL and the Si-face of 4H-SiC. A typical angle-resolved photoemission spectroscopy (ARPES) map on this system [14] is reported in Fig. 4.2b. Due to the covalent bond with the substrate, the π/π^* bands typical of sp^2 carbon cannot be observed. As schematically illustrated in Fig. 4.2c, the first graphene layer originated by the breaking of the covalent bonds with the Si-face, as a consequence of the formation of a new BL underneath the old one. Clearly, the ARPES spectrum of this system (see Fig. 4.2d) shows the typical linear dispersion relation of a single layer of graphene [14]. The shift of the Fermi level in the conduction band ($E_F - E_D \approx$ 0.45 eV) indicates a high n-type doping ($\sim 10^{13}$ cm^{-2}) of graphene residing on the BL. This has been ascribed to the charged dangling bonds at the interface between the BL and Si-face, which are also partially responsible for the lower mobility ($\sim 10^3$ cm^2/V s) [15] of a monolayer graphene on SiC(0001) with respect to the values ($\sim 10^4$ cm^2/V s) usually reported for graphene exfoliated from highly oriented pyrolytic graphite (HOPG) to common substrates [16–18].

An effective way to decouple the BL from (0001) face is the intercalation of proper atomic species, being hydrogen the most used one [14, 15]. Annealing in H_2 atmosphere breaks up the covalent bonds between the BL and SiC, saturates the dangling bonds, and converts the electrically inactive BL into an additional real graphene layer. As a result, a quasi-freestanding monolayer graphene (QFMLG) is obtained on the SiC surface covered by the BL, whereas a quasi-freestanding bilayer graphene (QFBLG) is obtained starting from the MLG residing on the BL. Although the passivation of Si dangling bonds removes the origin of high n-type doping in the as-grown epitaxial graphene on the Si-face, after hydrogen intercalation a residual hole doping ($p \approx 8 \times 10^{12}$ cm^{-2}) is still observed, which has been ascribed to the effect of polarization charge from the SiC substrate [19].

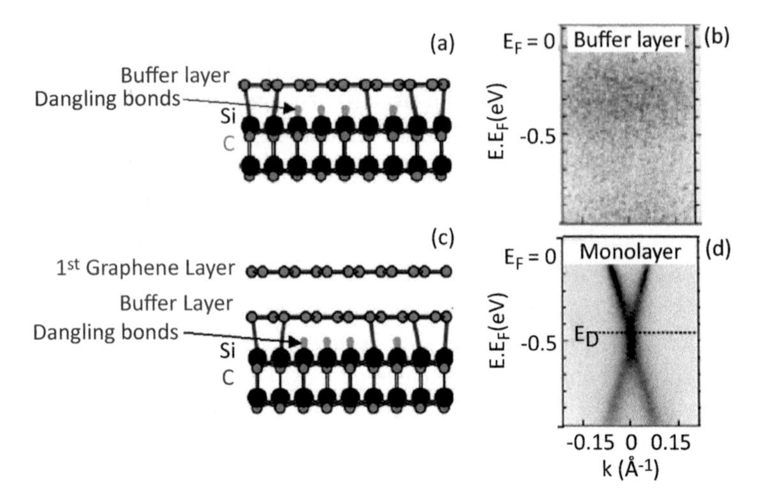

Figure 4.2 (a) Schematic representation of the interface between the buffer layer (BL) and the Si-face of 4H-SiC and (b) angle-resolved photoemission spectroscopy (ARPES) on this system. (c) Scheme of the BL and the first graphene layer and (d) ARPES spectrum of this system, showing a linear dispersion relation. Reprinted with permission from Ref. [14], Copyright 2009, American Physical Society.

As already mentioned, graphene films grown on (000-1) face typically consist of a laterally inhomogeneous distribution of rotationally misoriented multilayers. In spite of this structural disorder, the measured Hall effect mobilities on the C-face largely outscore those measured on the Si-face [20]. This is typically ascribed to the electronic decoupling between the different graphene layers on the C-face (behaving similarly to independent single layers) and to the electrostatic screening of inner layers in the stack from the external environment. However, the origin of these differences in the morphological and electrical properties of samples grown on the two polar SiC(0001) surfaces is still a matter of debate. Apart from a general acceptance of the absence of the BL and of a weaker interface coupling with respect to the Si-face [21], uncertainty still exists on whether graphene grows directly on the C-terminated surface [22] or if it is preceded by either ordered surface reconstructions [21] or a disordered interface layer [23, 24]. The role of substrate defects in the nucleation of graphene on the C-face has been also demonstrated [25]. Recent experimental investigations by aberration-corrected transmission electron microcopy and energy loss spectroscopy indicated for graphene grown on SiC(000-1) at high temperatures (1900°C) the presence of a thin (~1 nm) amorphous film at the interface, which strongly suppresses the epitaxy of graphene on the SiC substrate [26, 27]. This film has an almost fixed thickness regardless of the number of the overlying graphene layers, while its chemical signature shows the presence of C, Si, and O.

Interestingly, a BL-free graphene growth has been recently reported on (11–20) and (1–100) nonpolar SiC faces [11]. In particular, low-energy electron diffraction (LEED) patterns on few layers of graphene on the (11–20) face indicated the absence of rotational disorder between the layers, whereas ARPES measurements indicate that the Dirac point lies ~0.25 eV below the Fermi level, indicating an n-type doping in the order of 4×10^{12} cm^{-2} [11]. The origin of this doping is still unclear. A high degree of rotational disorder was, instead, observed by LEED in the case of graphene grown on the (1–100) face [11], similarly to the case of graphene grown on the (000-1) face.

4.3 Epitaxial Graphene on Steps and Facets of SiC(0001): Morphology and Structural Properties

Previous discussion showed that epitaxial graphene on (0001) plane differs from graphene grown on the other low-index crystallographic orientations of hexagonal SiC, due to the presence of a well-defined interfacial layer with the substrate, which determines both a high-quality epitaxial growth and the Bernal stacking between graphene layers.

As a matter of fact, for practical reasons, SiC wafers are always cut from ingots with angles different than zero with respect to the crystal growth axis [0001]. As a consequence, the Si-face of SiC wafers is composed by (0001) terraces separated by nanometric steps running perpendicularly to the miscut direction (typically [11–20] or [1–100]). For “nominally on-axis” wafers, i.e., with a very low unintentional miscut angle, the average terraces width can range from a few μm to tens of μm. In the case of off-axis SiC wafers, with typical miscut angles of 4° or 8°, the terrace width and, hence, the step distance are significantly smaller.

The wafer miscut angle plays an important role in the graphene growth process, since the kinks of SiC terraces are nucleation sites for epitaxial graphene. As a consequence of this, for fixed annealing conditions, a higher growth rate is observed on vicinal SiC (0001) surfaces than on "nominally on-axis ones" due to the reduced spacing between the terrace kinks [28]. Clearly, single-layer graphene coverage on a large SiC surface fraction can be easily achieved on on-axis substrates, whereas a few layers or multilayers of graphene are typically obtained on off-axis ones under similar growth conditions. A representative Raman spectrum of epitaxial graphene grown on nominally on-axis 4H-SiC(0001) is reported in Fig. 4.3a, along with the spectrum collected on the virgin substrate (for comparison) [29]. The spectra have been offset vertically for clarity and their intensity has been normalized to the intensity of the most prominent Raman peak of 4H-SiC substrate. Two main peaks are typical of the Raman spectrum of graphene films, i.e., the G peak at $\sim$1600 cm^{-1} and the 2D peak at $\sim$2690 cm^{-1} [30]. The line shape of the 2D peak and the intensity ratio between the 2D and G peaks provide information on the number of graphene layers and on the stacking order (in the case of few layers or multilayers). While the *G* peak at $\sim$1600 cm^{-1} is not easily distinguishable in the spectrum of Fig. 4.3a, being overlapped to the characteristic Raman features of the 4H-SiC substrate, the 2D peak at 2692 cm^{-1} is clearly evident. As better shown in the insert of Fig. 4.3a, this peak is symmetric and can be fitted by a single Lorentzian component with full width at half-maximum (FWHM) of $\sim$37 cm^{-1}, consistently with single-layer graphene coverage [31]. A typical AFM morphology of the same sample is reported in Fig. 4.3b, showing that graphene resides on a surface characterized by micrometer-wide terraces separated by nanometer-high steps. The height line profile in the direction [11–20] orthogonal to the terrace kinks (Fig. 4.3c) shows that nanosteps exhibit vertical sidewalls corresponding to (11–20) nonpolar faces of SiC.

A representative Raman spectrum on epitaxial graphene grown on a 4H-SiC(0001) substrate with 8° off-axis miscut angle along the [11–20] direction is reported in Fig. 4.4a. The typical G and 2D peaks of graphitic structures are clearly evident. The asymmetric shape of the 2D peak (insert) is consistent with the formation of a few layers of graphene with Bernal stacking [30]. A representative AFM image of 4H-SiC morphology after epitaxial graphene growth is reported in Fig. 4.4b, together with a line profile along the direction [11–20] (c). The AFM analysis shows a stepped surface with hundreds of nanometer-wide (0001) terraces alternating with (11-2n) facets. SiC steps run parallel to the [1–100] direction (i.e., perpendicularly to the wafer miscut direction) over the entire substrate. This morphology originates from the "step bunching" typically occurring in hexagonal SiC upon high-temperature annealing [28].

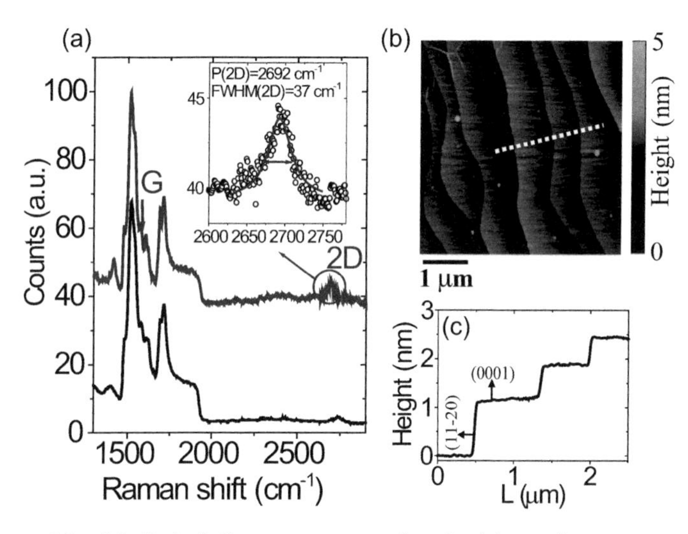

Figure 4.3 (a) Typical Raman spectra of epitaxial graphene grown on "nominally on-axis" 4H-SiC (0001) and of the virgin substrate. The 2D peak has been fitted with a single Lorentzian component with FWHM ≈37 cm^{-1} (see inset), indicating that the sample is mostly composed by single layers. (b) AFM morphology and (c) height line profile of the SiC surface covered by epitaxial graphene. Reprinted with permission from Ref. [29], Copyright 2012, American Physical Society.

Atomic resolution structural characterization of the interface between epitaxial graphene and the facets of SiC surface have been recently obtained using aberration-corrected scanning transmission electron microscopy (AC-STEM) [12]. Direct (atomic-scale) images on the sample cross section have been obtained by scanning an electron beam with Å (or even sub-Å) size and collecting transmitted electrons scattered at large angle (high-angle annular dark field, HAADF, operation mode). In order to avoid any artifact related to radiation damage of graphene during the measurement, a primary electron beam with energy of ~60 keV, i.e., below the knock-on threshold for carbon atoms in graphene (~80 keV) [32], was employed. At this energy, the estimated primary electron beam size is 1.1 Å.

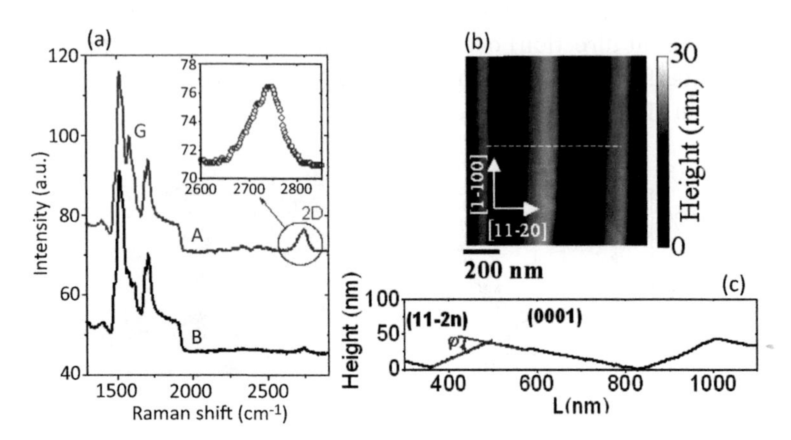

Figure 4.4 (a) Typical Raman spectra of epitaxial graphene grown on 8° off-axis 4H-SiC (0001), curve A, and of the virgin substrate, curve B. The asymmetric shape of the 2D peak (see insert) is consistent with the formation of few layers of graphene with Bernal stacking. AFM morphology (b) and height line profile (c) of the SiC surface after graphene growth. Reprinted with permission from Ref. [12], Copyright 2013, American Chemical Society.

Figure 4.5a shows an atomic resolution HAADF-STEM image of the graphene/4H-SiC interface acquired on a sample cross sectioned perpendicularly to the [1–100] direction. The (0001) basal plane and the (11-2n) facets form an angle ϕ, as illustrated in the inset of Fig. 4.5a. From statistics on different steps imaged by STEM, the value of this angle was found to be in the range from 27° to 34°. The graphene layers are clearly resolved at the atomic scale. This allows a precise measurement of the distances between the different carbon layers and the (0001) and (11-2n) SiC surfaces, as shown in the line scans of Fig. 4.5b,c. The separations between the topmost SiC surface dimer and the first C layer on the (0001) surface was 2.62 ± 0.13 Å, indicating a very strong bonding of this layer to the substrate, whereas the distance between the first and second carbon layers is 3.37 ± 0.13 Å, in close agreement with the typically reported interlayer spacing in graphite. By contrast, the distance between the (11-2n) surface and the first C layer above this facet was found to be 3.46 ± 0.13 Å, whereas the distance between the first and the second C layer was (3.37 ± 0.13 Å). This morphological difference was a direct (atomic-scale) evidence that the first C layer in the stack has a different nature depending on the orientation of the underlying substrate: It can be identified with the BL on the planar (0001) surface, whereas it appears to delaminate from the (11-2n) surface.

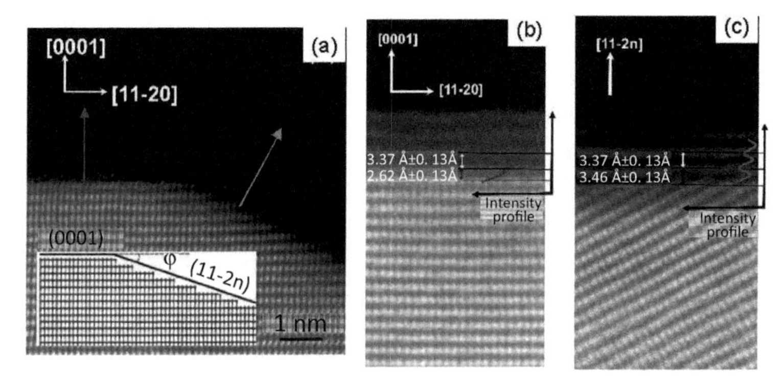

Figure 4.5 (a) Cross-sectional STEM image at atomic resolution of few layers of graphene in the kink region between basal plane (0001) and a facet (11-2n) of 4H-SiC. The image has been acquired at a 60 keV primary beam energy to prevent electron beam–induced damage to graphene. The insert reports a schematic of the (11-2n) SiC nanofacet structure, showing the angle φ between the (0001) plane and the (11-2n) facet. (b) STEM images acquired in the basal plane region. The intensity profile orthogonal to the (0001) surface is also reported, and the distances between SiC and the first C layer and the first and second C layers are indicated. (c) STEM image acquired in the facet region with the intensity profile orthogonal to the (11-2n) surface. The distances between this surface and the first C layer and the first and second C layers have been evaluated. Reprinted with permission from Ref. [12], Copyright 2013, American Chemical Society.

The results of this structural analysis were further supported by atomic-scale EELS. Figure 4.6a displays the EELS spectra of the carbon K edge fine structure performed on the first and second C layers residing both on the (0001) basal plane and on (11-2n) facets, as well as on the topmost SiC dimer (taken as reference). Marked differences appear in the fine structure of the carbon K edge depending on the layer for which the spectrum was acquired. The K edge of carbon typically comprises a sharp peak at 285 eV related to π^*-bonding. In the experimental geometry used here (where flat graphene layers are being observed by STEM in cross section), it can be directly associated with the presence of sp^2 hybridization, as it is expected for graphene. For instance, the π^* peak is quite prominent on the second C layer of the planar (0001) surface, while its intensity drops significantly at the first C layer of the same surface, i.e., at the BL. The reduction of the intensity of π^*-bonding at the BL is accompanied by a simultaneous increase in σ^*-bonding, seen through the increasing intensity of the second peak in the EELS K edge at ~295 eV, which indicates the presence of sp^3 hybridization. By contrast, when following the first C layer from the (0001) surface to the (11-2n) facet, the π^* peak intensity at 285 eV increases to a value comparable to what is measured on the outer graphene layers. This suggests the absence of any sp^3 hybridization in the first layer above the SiC(11-2n) facet; hence, it can no longer be described as a BL but as a pure graphene-like layer on the top of the surface facets.

The observed BL delamination has been explained in terms of a lowering of the graphene binding energy with the substrate moving from the (0001) basal plane to (11-2n) facets. Basing on energetic considerations, it has been argued that this delamination occurs above a critical angle between (11-2n) and (0001) planes, which was estimated as $\varphi \geq 27°$ [12].

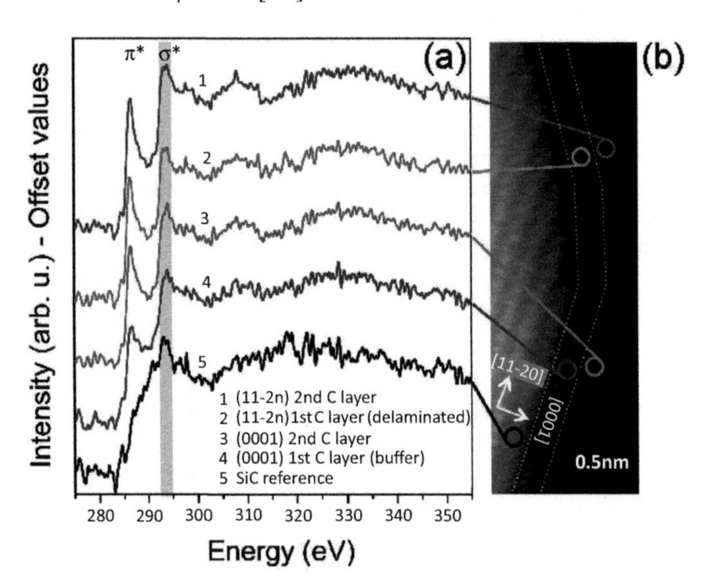

Figure 4.6 (a) EELS spectra acquired on different position on the first and second C layers residing on the (0001) basal plane and (11-2n) facets and on the SiC topmost dimer (reference), as indicated in the cross-sectional STEM image in (b). Reprinted with permission from Ref. [12], Copyright 2013, American Chemical Society.

As discussed in the next section, this difference in the interface structure between graphene on the SiC basal plane and on facets has strong implications in the electrical behavior of this system. Since the high average n-type doping of epitaxial graphene on SiC(0001) is typically ascribed to the presence of the BL, the local BL delamination from facets is expected to lead to a locally reduced carrier density. Such hypothesis is confirmed by local electrical measurements by conductive AFM, which show a local increase in the graphene resistance on facets with respect to the basal plane.

4.4 Electronic Properties of Epitaxial Graphene Residing on Nanosteps and Facets

4.4.1 Anisotropic Current Transport in Epitaxial Graphene Devices

The aforementioned morphological and structural properties of epitaxial graphene residing on nanosteps and facets of hexagonal SiC, Si-face, have a significant impact on the overall electrical properties of epitaxial graphene–based devices. A commonly observed phenomenon is an anisotropy of current transport within epitaxial graphene depending on the current flow direction with respect to the orientation of SiC steps or facets. A reduced conductance is typically found in the direction perpendicular to steps and facets with respect to that measured in the parallel direction. Such an effect has been reported by different kinds of measurements, such as four-point probe [33], transmission line model (TLM) [12, 34], field-effect transistors (FETs) [35, 36], and quantum Hall effect measurements on Hall bars [37, 38].

As an example, Fig. 4.7a shows the optical microscopy of linear TLM test patterns fabricated on few layers of graphene samples grown on 8°-off SiC(0001) wafers. These structures, consisting of a set of Ni/Au Ohmic contacts at different spacings (from 20 to 100 μm) on laterally isolated graphene rectangular regions, were fabricated along two different orientations, i.e., with the direction of current flow parallel or perpendicular to the substrate steps.

The resistance between adjacent pads as a function of the pad distance is reported in Fig. 4.7b, clearly showing an asymmetry in the macroscopic current flow along the two directions. The sheet resistances of the few layers of graphene along the two directions, evaluated from the slope of the linear fit for the two sets of data, were 775 ± 28 Ω/sq in the [11−20] direction and 397 ± 20 Ω/sq in the [1−100] direction.

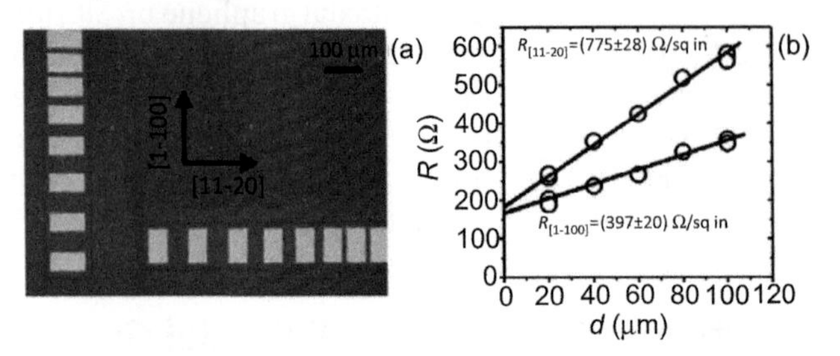

Figure 4.7 (a) Optical microscopy of TLM test patterns oriented in the direction parallel ([1−100]) and orthogonal [11−20] to the substrate steps. (b) Measured resistance (*R*) versus the contacts distance (*d*) in the two directions. The evaluated sheet resistances from the fit of experimental data are also reported. Reprinted from Ref. [41], with permission from Trans Tech Publications.

An anisotropy of channel conductance was also observed in few layers of graphene FETs fabricated with the channel length parallel or perpendicular to the substrate steps. Figure 4.8a shows a cross-sectional schematic and Fig. 4.8b an optical microscopy of the top-gated FET structures with an SiN gate dielectric (thickness $t_{\text{SiN}} \sim 40$ nm) deposited by plasma-enhanced chemical vapor deposition (PECVD). Long-channel devices (gate length L_G from 5 to 20 μm and channel width $W = 100$ μm) were used to maximize the importance of the channel resistance with respect to the contact resistance on the source–drain current.

Figure 4.8c shows the channel conductance $g_D = [dI_d/dV_d]V_{tg}$ versus top gate bias (V_{tg}) of FETs with the channel length in the direction parallel ([1−100]) and perpendicular ([11−20]) to substrate steps. For both channel orientations, the g_D–V_{tg} characteristics exhibit an ambipolar behavior, with the minimum conductivity (neutrality point) shifted to negative gate bias ($V_{NP} \approx -10$ V), consistently with the average n-type doping of epitaxial graphene. The average carrier density (n) as a function of the gate bias (upper scale in Fig. 4.8c) was calculated as $n = \varepsilon_0\varepsilon(V_{tg} - V_{NP})/(qt_{\text{SiN}})$, where ε_0 and $\varepsilon \approx 7$ are the vacuum dielectric constant and the SiN relative dielectric constants, respectively, and q is the electron charge. A significantly higher channel conductance is measured in the [1−100] direction with respect to [11−20] direction. As an example, at $V_{tg} = 0$, $g_{D[1-100]} \approx 2.3$ mS, i.e., ~77% higher than

$g_{D[11-20]} \approx 1.3$ mS. The effective channel mobility μ in the electron branch was also evaluated as $\mu = g_d/(qn)$ and is depicted in Fig. 4.8d. The mobility values for the two orientations are very similar (~2900 cm²/Vs) at V_{tg} approaching the neutrality point, whereas $\mu_{[11-20]} \approx 830$ cm²/Vs and $\mu_{[1-100]} \approx 1470$ cm²/Vs were obtained at $V_{tg} = 0$ V.

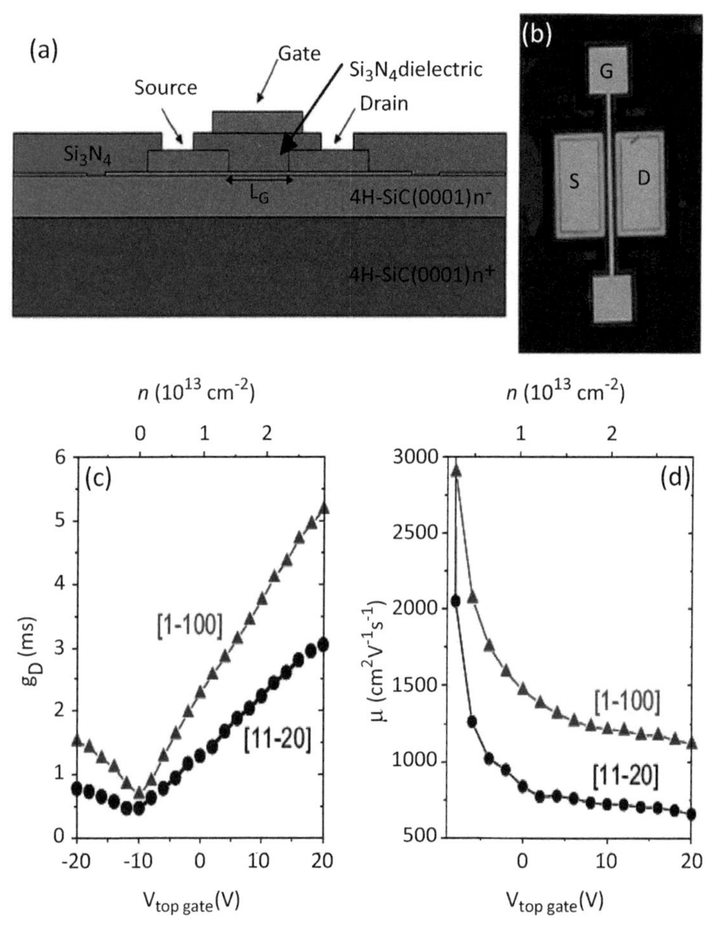

Figure 4.8 Cross-sectional schematics (a) and optical microscopy (b) of a top-gated epitaxial graphene field-effect transistor (FET) with channel length L_G = 10 μm and channel width W = 100 μm. (c) Channel conductance versus top gate bias and carrier density for FETs with the channel length along the [11–20] direction, i.e., orthogonal to the SiC steps, and the [1–100] direction, i.e., parallel to the SiC steps. (d) Evaluated field-effect mobility μ in the electron branch for FETs with the channel length along [11–20] and [1–100] directions. Reprinted from Ref. [36], Copyright 2014, with permission from Elsevier.

4.4.2 Nanoscale Resistance Measurements

The observed current transport anisotropy in epitaxial graphene devices on stepped or faceted SiC clearly indicates the occurrence of different resistive contributions for current flowing in the direction parallel and perpendicular to the steps/facets.

In the last years, several investigations have been carried by different research groups using scanning probe techniques, such as scanning probe potentiometry [29] and conductive atomic force microscopy (CAFM) [29, 34, 40], in order to elucidate the origin of such electrical anisotropy. The result of these investigations was the observation of a local increase in the resistance of epitaxial graphene over nanosteps or facets with respect to the case of the basal plane.

In the following is illustrated an approach based on the use of CAFM and properly designed test patterns, which allowed to quantitatively evaluate the local resistance (R_{loc}) in epitaxial graphene. An optical microscopy of the used structure, fabricated by metal lift off on epitaxial graphene, is reported in Fig. 4.9a. This test pattern is formed by circular metal contacts (indicated with numbers from 1 to 5) with inner radius r_2 = 200 μm separated by metal-uncoated graphene annular rings of variable widths (*d* from 5 to 25 μm) from the outer metal contact. A detail of these structures is shown in the optical microscopy in Fig. 4.9b. This kind of electrical test structures is typically named circular TLM. The pattern contains also a metal-uncoated circular graphene area with radius r_2 = 200 μm, on which nanoscale current measurements by CAFM were carried out (see schematic in Fig. 4.9c).

CAFM scans were performed by metal-coated AFM tips on ~1 μm × 1 μm regions close to the center of that circular area. The lateral current flow inside graphene between the macroscopic contact and the nanometric tip (with curvature radius $r_{tip} \approx 20$ nm) was measured by a current amplifier connected to the tip. In this configuration, the measured resistance *R* can be expressed as the series, i.e., the sum of different contributions (see schematic in Fig. 4.9d):

$$R = R_{c,tip} + R_{spr} + R_{series} + R_{c,macro} \quad (4.1)$$

with $R_{c,tip}$ the resistance of the nanometric tip contact onto graphene, R_{spr} the spreading resistance encountered by the current to spread from the tip contact within graphene, R_{series} the series resistance, and $R_{c,macro}$ the contact resistance of the macroscopic Ohmic contact. It is worth noting that, while the $R_{c,tip}$ and $R_{c,macro}$ contributions are associated to the vertical current transport from the tip to graphene and from graphene to the macroscopic electrode, respectively, the R_{spr} and R_{series} contributions are associated to horizontal current transport inside graphene. $R_{c,tip}$ is due to the current tunneling through the Schottky barrier between the tip and graphene and can be minimized by properly setting the contact force between the tip and graphene. R_{spr} captures the information on the "local" resistance R_{loc} of graphene with a lateral resolution comparable with the tip radius. Due to the radial symmetry of the system and considering the 2D nature of epitaxial graphene, R_{spr} can be expressed as:

$$R_{spr} = \frac{R_{loc}}{2\pi} \ln\left(\frac{r_{tip}}{r_0}\right), \tag{4.2}$$

where the radial distance r_0 can be assumed as $r_0 \approx 0.33$ nm, i.e., the distance of 1L of graphene from the BL. The series resistance contribution is a constant value related to the average sheet resistance (R_{av}) of graphene on an mm^2 area and can be expressed as:

$$R_{series} = \frac{R_{av}}{2\pi} \ln\left(\frac{r_2}{r_{tip}}\right) \tag{4.3}$$

Both the average sheet resistance R_{av} and the macroscopic contact resistance $R_{c,macro}$ have been independently evaluated by the electrical characterization of the circular TLM structures [29]. This method allows to evaluate R_{loc} of epitaxial graphene at nanometer scale and to correlate the R_{loc} values with surface morphology.

The following paragraphs illustrate some applications of this method to relevant case studies for epitaxial graphene, i.e., (i) the case of a graphene monolayer over SiC vertical nanosteps, (ii) the case of monolayer/bilayer lateral junctions, and (iii) the case of graphene residing on oblique SiC facets.

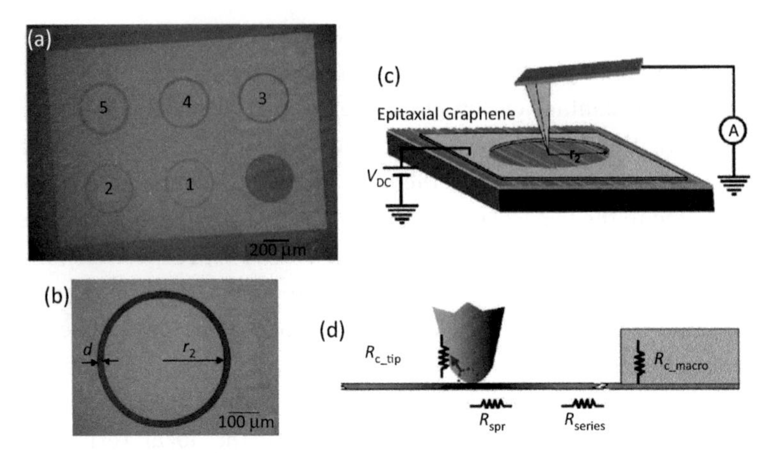

Figure 4.9 (a) Optical microscopy of test patterns for nanoscale and macroscopic electrical characterization of epitaxial graphene on SiC(0001). (b) Detail of the test structure #3. (c) Schematics of the experimental setup for CAFM measurements. (d) Representation of the different resistance contributions to the measured resistance *R* by CAFM. Reprinted with permission from Ref. [29], Copyright 2012, American Physical Society.

4.4.2.1 Monolayer graphene over nanosteps

Figure 4.10 shows the surface morphology (a) and the conductance map (b) in the case of a monolayer (1L) of graphene over a nanometric step (height $h \sim 1.5$ nm) of "nominally on-axis" SiC. The line profiles of the height (c) and of the evaluated local resistance R_{loc} (d) for the selected areas in the maps are also reported. A 55% enhancement of R_{loc} has been evaluated for graphene on the (11–20) vertical step, as compared to graphene on the surrounding (0001) planar regions. By performing a statistics over several SiC steps of different heights, the ratio between the local resistance of 1L graphene measured on the steps and on the terraces was found to increase as a function of the step height, as illustrated in Fig. 4.10e.

This trend is in agreement with what is reported by other authors [40] and suggests that this effect is related to the interaction between graphene and the sidewall face of the SiC step, most probably to a reduced electrostatic doping of graphene by the steps nonpolar face (11–20). This explanation would be also consistent with the increase in the resistance with the step height.

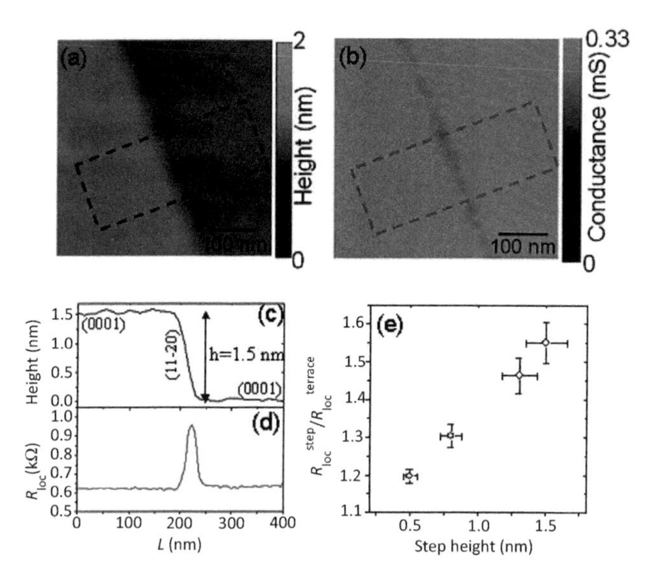

Figure 4.10 (a) AFM morphology and (b) conductance map on 1L graphene over a 1.5 nm substrate step. (c) Height profile extracted from the selected area on the morphology map. (d) Profile of the graphene local resistance calculated from the conductance map. (e) Ratio of the local resistance in 1L graphene over a substrate step and on the planar region as a function of the step height. Reprinted with permission from Ref. [29], Copyright 2012, American Physical Society.

4.4.2.2 Monolayer/bilayer graphene junctions

Figure 4.11a shows the conductance map measured on a representative area with submicrometer un-uniformities in the layer number, i.e., 2L, 1L, and the BL. A line profile of the local resistance R_{loc} in the 1L/2L junction region is also reported in Fig. 4.11b.

$R_{loc}(1L) \approx 0.62$ kΩ and $R_{loc}(2L) \approx 0.30$ kΩ have been obtained on 1L and 2L regions, respectively. A remarkable increase in the local resistance is observed at the 1L/2L junction with respect to the value measured in the 1L region [$R_{loc}(1L/2L) \approx 1.5R_{loc}(1L)$]. It is worth noting that this peak is very sharp and its width is just limited by the lateral resolution of the measurement, almost coincident with the tip-contact radius (~20 nm), suggesting that the effect responsible for the increased conductance is extremely localized.

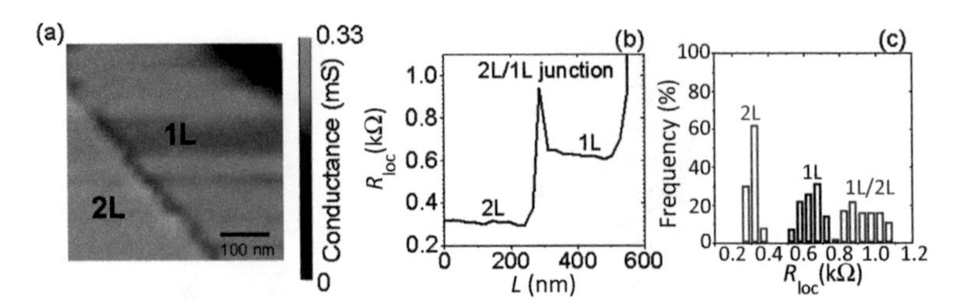

Figure 4.11 (a) Conductance map on a sample region with 2 layers (2L), 1 layer (1L), and the BL. (b) Local resistance profile (R_{loc}) in the 1L/2L junction region. (c) Histograms of the R_{loc} measured on 1L and 2L graphene regions and at the 1L/2L junctions in 10 different sample positions. Reprinted with permission from Ref. [29], Copyright 2012, American Physical Society.

This localized resistance enhancement has been consistently observed over different 1L/2L junctions identified in various sample positions by a cross-comparison of morphological and conductance maps. In Fig. 4.11c, the histograms of the R_{loc} values measured in 10 different positions on 1L and 2L plateau regions and at the respective interfaces have been reported. By fitting the three histograms with Gaussian distributions, the following average values of R_{loc} and root mean square standard deviations have been found:

$R_{loc}(1L) = (640 \pm 136)\ \Omega$, $R_{loc}(2L) = (313 \pm 60)\ \Omega$,
and $R_{loc}(1L/2L) = (928 \pm 210)\ \Omega$.

Noteworthy, the distribution of $R_{loc}(1L/2L)$ values is significantly larger than the corresponding $R_{loc}(1L)$ and $R_{loc}(2L)$ distributions. Although the enhancement of R_{loc} at the 1L/2L junction is in the same order of magnitude of that observed for 1L graphene over substrate steps, the origin of the two phenomena is expected to be completely different.

The first-principles quantum transport calculations have been applied to investigate the origin of the conductance degradation observed experimentally at the 1L/2L boundaries of epitaxial graphene on SiC. Figure 4.12a illustrates the system used for these calculations, consisting of a 1L graphene channel of length $d_{1L} = 8.6$ nm between two bilayer (2L) semi-infinite regions. Bernal stacking was considered in the 2L regions, and an interface with zigzag edges was considered at the 1L/2L boundary. The two bilayer regions were assumed to be contacted by ideally transparent electrodes, in such a way that all scattering mechanisms were due to the presence of the 1L/2L junction. Figure 4.12b shows the calculated ballistic conductance G (in units of $G_0 = 2q^2/h$, with q the electron charge and h the Planck's constant) as a function of the electron energy for the above-illustrated system with 1L/2L junctions (red symbols). For comparison, the calculated conductance for homogeneous 1L (black symbols) and 2L (blue symbols) systems with the same length have also been reported. This comparison shows a strong reduction of G in the presence of 1L/2L junctions for energies within a ±0.48 eV range from the Dirac point. Figure 4.12c shows a false color plot of the calculated transmission coefficient through the 1L/2L junction as a function of the electron energy and wave vector **k** within the first Brillouin zone. For comparison, the calculated transmission coefficients for uniform 1L and 2L graphene have also been reported in the insert. This plot shows a strong suppression of the transmittance around the Dirac point due to a destructive interference between the electron wavefunctions of monolayer's π/π^* bands and the wavefunctions of the bilayer's first bands. Hence, transport calculations in the ballistic regime suggest that conductance degradation in the presence of 1L/2L junctions can be ascribed to quantum interference effects.

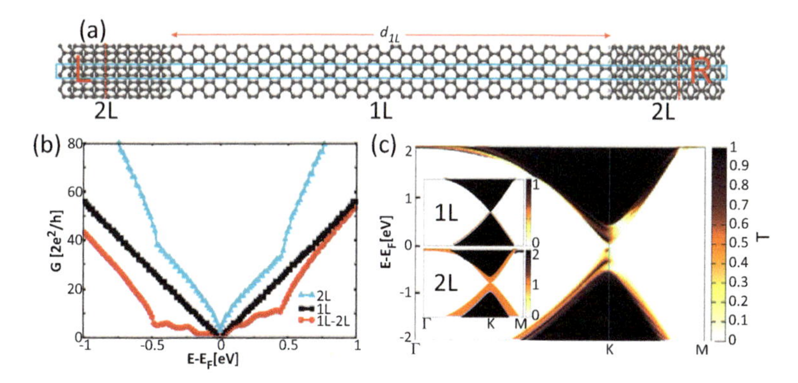

Figure 4.12 Scheme of the 1L–2L graphene double-junction system with a zigzag interface. L and R denote the positions of the left and the right contact, respectively. (b) Calculated conductance (in units of $G_0 = 2q^2/h$) as a function of energy for the 1L/2L junction (circles) and 1L graphene (squares), 2L graphene (triangles). (c) Momentum–space representation of the transmission coefficient as a function of energy for the 1L/2L junction. Insets show the respective transmission coefficients for ideal 1L and 2L graphene. Reprinted with permission from Ref. [29], Copyright 2012, American Physical Society.

4.4.2.3 Few layers of graphene on facets

An enhancement in the local resistance has also been observed in the case of few layers of graphene residing on (11-2n) facets of vicinal SiC surfaces with respect to that measured on the basal plane (0001).

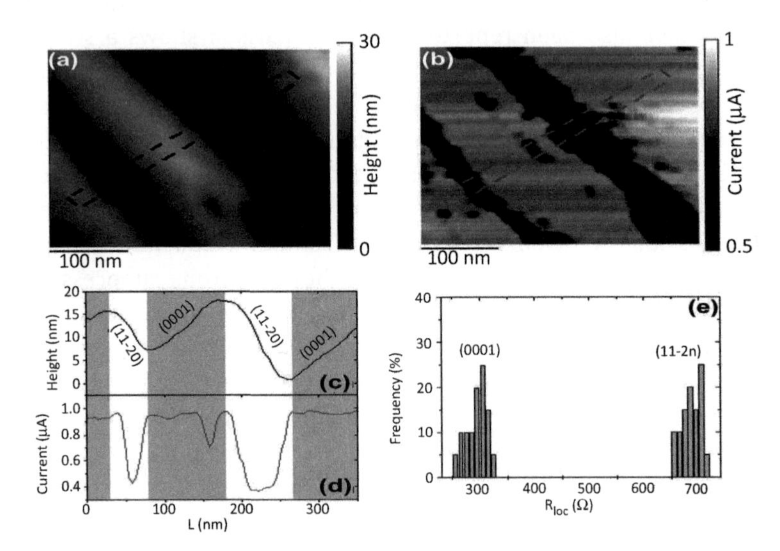

Figure 4.13 Surface morphology (a) and the corresponding current map (b) measured by CAFM on epitaxial graphene grown on 4H-SiC, Si-face, with 8° miscut angle in the [11-2n] direction. Line profiles of the height (c) and of the current (d) along the indicated directions in the maps. (e) Histograms of local resistances R_{loc} evaluated on epitaxial graphene residing on (0001) basal plane terraces and on facets (11-2n). Reprinted with permission from Ref. [12], Copyright 2013, American Chemical Society.

As an example, Fig. 4.13 shows the surface morphology (a) and the corresponding current map (b) acquired by CAFM in the case of few layers of epitaxial graphene grown on a 4H-SiC(0001) substrate with 8° miscut angle along the [11–20] direction. The line profiles of the height and of the current along the indicated directions in the maps are also depicted in Figs. 4.13c and 4.13d, respectively. A significant decrease (from ~1 to ~0.4 μA, i.e., more than a factor of two) is observed for the current measured on the (11-2n) facets with respect to that measured on the (0001) terraces. This is clearly correlated to an increase in the local resistance R_{loc} of few layers of graphene over these facets. Finally, in Fig. 4.13e, the histograms of the R_{loc} values evaluated on several different terraces and facets within the sample are reported. The average value and the standard deviation for the two distributions were extracted by Gaussian fitting of the histograms, obtaining $R_{loc} = (300 \pm 18)\ \Omega$ on the terraces and $R_{loc} = (688 \pm 24)\ \Omega$ on facets.

The local increase in graphene resistance on facets is fully coherent with the observed BL's delamination from these substrate regions and can be partially ascribed to a locally reduced doping of epitaxial graphene [12].

4.4.3 Correlating Macroscopic Current Transport Anisotropy with Nanoscale Resistance Inhomogeneities

These nanoscale observations explain the observed macroscopic current transport anisotropy in epitaxial graphene–based devices. In the following, a simple model accounting for the observed phenomena is presented [41].

As schematically illustrated in Fig. 4.14a, the average sheet resistance $R_{[11-20]}$ measured in the direction perpendicular to substrate steps can be expressed as the series combination of the local resistances on (0001) terraces ($R_{T,i}$) and of the local resistances on (11-2n) facets ($R_{F,i}$):

$$R_{[11-20]} = \frac{\sum_i R_{T,i} L_{T,i} + \sum_i R_{F,i} L_{F,i}}{\sum_i L_{T,i} + \sum_i L_{F,i}} \approx \frac{\langle L_T \rangle \langle R_T \rangle + \langle L_F \rangle \langle R_F \rangle}{\langle L_T \rangle + \langle L_F \rangle} \quad (4.4)$$

with $L_{T,i}$ and $L_{F,i}$ the widths of the terraces and facets, respectively. In the second term of Eq. (4.4), $\langle L_T \rangle$ and $\langle R_T \rangle$ represent the average of the $L_{T,i}$ and $R_{T,i}$ values over a large number of terraces, while $\langle L_F \rangle$ and $\langle R_F \rangle$ are the average of the $L_{F,i}$ and $R_{F,i}$ values over a large number of facets. On the other hand, as illustrated in Fig. 4.14b, the average sheet resistance $R_{[1-100]}$ measured in the direction parallel to the steps can be expressed as the parallel combination of $R_{T,i}$ and $R_{F,i}$:

$$R_{[1-100]} = \frac{\sum_i L_{T,i} + \sum_i L_{F,i}}{\sum_i \frac{L_{T,i}}{R_{T,i}} + \sum_i \frac{L_{F,i}}{R_{F,i}}} \approx \frac{\langle L_T \rangle + \langle L_F \rangle}{\frac{\langle L_T \rangle}{\langle R_T \rangle} + \frac{\langle L_F \rangle}{\langle R_F \rangle}}. \quad (4.5)$$

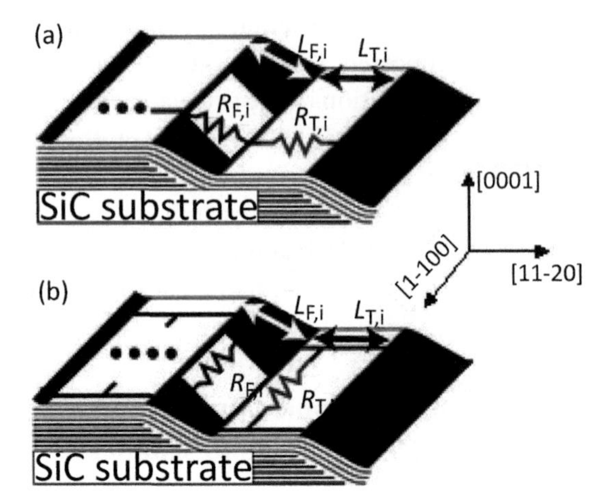

Figure 4.14 Schematic representation of the resistive contributions associated to terraces and facets for current transport in the directions orthogonal, i.e., [11–20], and parallel, i.e., [1–100], to SiC steps. Reprinted from Ref. [41], with permission from Trans Tech Publications.

As a matter of fact, the equivalent sheet resistance $R_{[11\text{-}20]}$ resulting from the series combination in Eq. (4.4) is higher than the equivalent sheet resistance $R_{[1\text{-}100]}$ resulting from the parallel combination in Eq. (4.5).

4.4.4 Local Electron Mean Free Path

As discussed in the previous section, the local increase in graphene resistance on SiC facets can be correlated to a local decrease in graphene carrier density, due to the BL local delamination from facets. However, an effect of the missing BL on the "local" transport properties, i.e., on the electron mean free path, of graphene residing on facets cannot be excluded.

Recent reports focused on the electrical characterization of epitaxial graphene nanoribbons produced by selective growth on SiC facets during thermal annealing of a pre-structured SiC(0001) substrate [42]. Transport measurements revealed the occurrence of ballistic transport in these graphene nanostructures [43]. The absence of the BL on facets was also invoked by the authors as a possible structural justification of these excellent electronic properties [43].

Local measurements of the electron mean free path in epitaxial graphene residing on a faceted SiC surface have also been recently performed using a scanning probe method based on local capacitance measurements. The principles of this technique are widely discussed in Refs. [17, 18, 44].

The scanning capacitance microscopy (SCM) setup used to perform local capacitance measurements [45–47] is schematically illustrated in Fig. 4.15a. The sample surface was scanned in contact mode using a metal-coated AFM tip, and a DC bias ramp was applied to the sample back gate to sweep the Fermi level in graphene from the hole to the electron branch. For the sample in this study, the back gate is provided by the n^+-SiC substrate under the n^--epi-SiC layer. In addition, an AC modulating bias is superimposed to the DC bias to induce an excess of charges from the nanometric tip contact to graphene. This charge in excess diffuses in graphene over an effective area A_{eff} around the tip contact, whose extension depends on the gate bias. An ultrahigh-sensitive capacitance sensor is connected to the tip and measures the capacitance variations induced by the modulating bias, using a lock-in method. The capacitance measured when the tip is in contact with graphene (C_{gr}) is compared to the capacitance measured on and identical SiC sample without graphene (C_{SiC}). In the latter case, C_{SiC} is the capacitance of a parallel plate capacitor with area A_{tip}, the tip-contact area, i.e., $C_{\text{SiC}} = A_{\text{tip}}\varepsilon_0\varepsilon_{\text{SiC}}/t_{\text{SiC}}$, where ε_0 is the vacuum permittivity, ε_{SiC} is the SiC permittivity, and t_{SiC} is the SiC epilayer thickness. Instead, C_{gr} is the capacitance of a parallel plate capacitor with area A_{eff}, i.e., $C_{\text{gr}} = A_{\text{tip}}\varepsilon_0\varepsilon_{\text{SiC}}/t_{\text{SiC}}$. For each tip position on graphene surface, A_{eff} is evaluated as $A_{\text{eff}} = A_{\text{tip}}C_{\text{gr}}/C_{\text{sub}}$. As previously pointed out, A_{eff} is the effective area where free carriers diffuse from the tip/graphene point contact. In the diffusive transport regime, it is related at room temperature to the "local" electron mean free path l_{loc} in graphene as $l_{\text{loc}} = (A_{\text{eff}}/\pi)^{1/2}$ [44].

Figure 4.15b shows the surface morphology of the investigated epitaxial graphene residing on the faceted SiC surface. As illustrated in the same image, the tip is displaced on a regular array of positions with ~100 nm spacing and for each position, the local electron mean free path (l_{loc}) was measured. Clearly, this array includes both terraces and facets regions. In Fig. 4.15c, the histogram of the probed l_{loc} values is reported. A distribution ranging from ~10 to ~120 nm can be observed, with two main peaks (at ~30 and ~100 nm), which are associated to the mean free path values on (0001) terraces and (11-2n) facets, respectively. This latter measurement is a direct evidence of a significant improvement (~3×) of transport properties in BL-free epitaxial graphene on facets, probably due to a strong reduction in Coulomb scattering of graphene electrons.

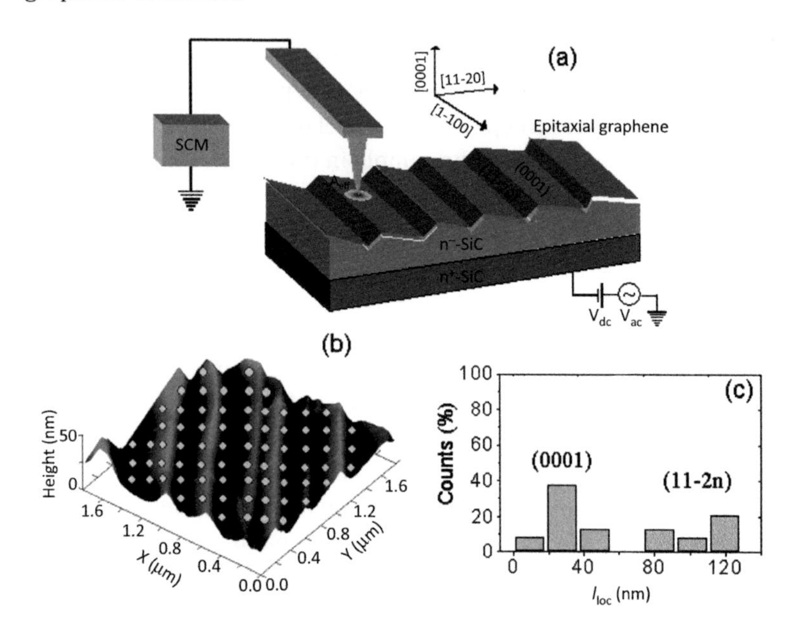

Figure 4.15 (a) Schematic representation of the experimental setup for local electron mean free path measurements based on scanning capacitance microscopy. (b) AFM morphology of the investigated epitaxial graphene grown on SiC, Si-face, 8° off. The array of tip positions where local mean free path was measured is also indicated. (c) Histogram of the local mean free path values (l_{loc}) for graphene on (0001) terraces and (11-2n) facets. Reprinted from Ref. [36], Copyright 2014, with permission from Elsevier.

4.5 Summary

In summary, this chapter presented an overview of recent investigations of the morphological, structural, and electrical properties of epitaxial graphene residing on the stepped/faceted surface of vicinal SiC(0001) wafers. The combination of atomic resolution structural analyses by AC-STEM with nanoscale resolution electrical characterization based on AFM elucidated the key role played by the steps/facets of the SiC substrate on the local electrical behavior of epitaxial graphene. These nanoscale electrical analyses have been correlated with the results of the macroscopic electrical characterization of epitaxial graphene device structures, providing the explanation of typically observed effects, such as the anisotropic current transport with respect to SiC steps orientation. Finally, local transport measurements showed a significantly higher electron mean free path for graphene residing on (11-2n) SiC facets than on the (0001) basal plane, consistently with recent reports on exceptional ballistic transport in epitaxial graphene nanoribbons.

Acknowledgments

The authors want to acknowledge S. Di Franco (CNR-IMM) for the support in graphene devices fabrication, G. Fisichella (CNR-IMM) for AFM measurements, P. Fiorenza and G. Greco (CNR-IMM) for device's electrical characterization, R. Lo Nigro (CNR-IMM) for dielectrics depositions. N. Piluso (CNR-IMM), A. Piazza (CNR-IMM and University of Catania) and S. Agnello (University of Palermo) for Raman analyses, C. Bongiorno and M. Scuderi (CNR-IMM), P. Longo (Gatan Inc.), Q. Ramasse (SuperSTEM Laboratory, Daresbury, UK) for the collaboration in STEM experiments. This work has been funded, in part, by the FlagERA project GraNitE.

References

1. Berger, C., Song, Z., Li, X., Wu, X., Brown, N., Naud, C., Mayou, D., Li, T., Hass, J., Marchenkov, A. N., Conrad, E. H., First, P. N., and de Heer, W. A. (2006). Electronic confinement and coherence in patterned epitaxial graphene, *Science*, **312**, pp. 1191–1196.
2. Emtsev, K. V., Bostwick, A., Horn, K., Jobst, J., Kellogg, G. L., Ley, L., McChesney, J. L., Ohta, T., Reshanov, S. A., Rohrl, J., Rotenberg, E., Schmid, A. K., Waldmann, D., Weber, H. B., and Seyller, T. (2009). Towards wafer-size graphene layers by atmospheric pressure graphitization of SiC(0001), *Nat. Mater.*, **8**, pp. 203–207.
3. Virojanadara, C., Syvajarvi, M., Yakimova, R., Johansson, L. I., Zakharov, A. A., and Balasubramanian, T. (2008). Homogeneous large-area graphene layer growth on 6H-SiC(0001), *Phys. Rev. B*, **78**, 245403.

4. Moon, J. S., Curtis, D., Hu, M., Wong, D., McGuire, C., Campbell, P. M., Jernigan, G., Tedesco, J. L., VanMil, B., Myers-Ward, R., Eddy, Jr., C., and Gaskill, D. K. (2009). Epitaxial-graphene RF field-effect transistors on Si-face 6H-SiC substrates, *IEEE Electron Device Lett.*, **30**, pp. 650–652.

5. Lin, Y.-M., Dimitrakopoulos, C., Jenkins, K. A., Farmer, D. B., Chiu, H.-Y., Grill, A., and Avouris, Ph. (2010). 100-GHz transistors from wafer-scale epitaxial graphene, *Science*, **327**, p. 662.

6. Tzalenchuk, A., Lara-Avila, S., Kalaboukhov, A., Paolillo, S., Syväjärvi, M., Yakimova, R., Kazakova, O., Janssen, T. J. B. M., Fal'ko, V., and Kubatkin, S. (2010). Towards a quantum resistance standard based on epitaxial graphene, *Nat. Nanotechnol.*, **5**, pp. 186–189.

7. Pearce, R., Iakimov, T., Andersson, M., Hultman, L., Lloyd Spetz, A., and Yakimova, R. (2011). Epitaxially grown graphene-based gas sensors for ultrasensitive NO_2 detection, *Sens. Actuators B*, **155**, pp. 451–455.

8. Suemitsu, M., Jiao, S., Fukidome, H., Tateno, Y., Makabe, I., and Nakabayashi, T. (2014). Epitaxial graphene formation on 3C-SiC/Si thin films, *J. Phys. D Appl. Phys.*, **47**, 094016 (11pp).

9. Emtsev, K. V., Speck, F., Seyller, T., Ley, L., and Riley, J. D. (2008). Interaction, growth, and ordering of epitaxial graphene on SiC{0001} surfaces: A comparative photoelectron spectroscopy study, *Phys. Rev. B*, **77**, 155303.

10. Jabakhanji, B., Camara, N., Caboni, A., Consejo, C., Jouault, B., Godignon, P., and Camassel, J. (2012). Almost free standing graphene on SiC(000-1) and SiC(11-20), *Mater. Sci. Forum*, **711**, pp. 235–241.

11. Ostler, M., Deretzis, I., Mammadov, S., Giannazzo, F., Nicotra, G., Spinella, C., Seyller, Th., and La Magna, A. (2013). Direct growth of quasi-free-standing epitaxial graphene on nonpolar SiC surfaces, *Phys. Rev. B*, **88**, 085408.

12. Nicotra, G., Ramasse, Q. M., Deretzis, I., La Magna, A., Spinella, C., and Giannazzo, F. (2013). Delaminated graphene at silicon carbide facets: Atomic scale imaging and spectroscopy, *ACS Nano*, **7**, pp. 3045–3052.

13. Varchon, F., Feng, R., Hass, J., Li, X., Ngoc Nguyen, B., Naud, C., Mallet, P., Veuillen, J.-Y., Berger, C., Conrad, E. H., and Magaud, L. (2007). Electronic structure of epitaxial graphene layers on SiC: Effect of the substrate, *Phys. Rev. Lett.*, **99**, 126805.

14. Riedl, C., Coletti, C., Iwasaki, T., Zakharov, A. A., and Starke, U. (2009). Quasi-free-standing epitaxial graphene on SiC obtained by hydrogen intercalation, *Phys. Rev. Lett.*, **103**, 246804.

15. Speck, F., Jobst, J., Fromm, F., Ostler, M., Waldmann, D., Hundhausen,

M., Weber, H., and Seyller, T. (2011). The quasi-free-standing nature of graphene on H-saturated SiC(0001), *Appl. Phys. Lett.*, **99**, 122106.

16. Ni, Z. H., Ponomarenko, L. A., Nair, R. R., Yang, R., Anissimova, S., Grigorieva, I. V., Schedin, F., Blake, P., Shen, Z. X., Hill, E. H., Novoselov, K. S., and Geim, A. K. (2010). On resonant scatterers as a factor limiting carrier mobility in graphene, *Nano Lett.*, **10**, pp. 3868–3872.
17. Sonde, S., Giannazzo, F., Vecchio, C., Yakimova, R., Rimini, E., and Raineri, V. (2010). Role of graphene/substrate interface on the local transport properties of the two-dimensional electron gas, *Appl. Phys. Lett.*, **97**, 132101.
18. Giannazzo, F., Sonde, S., Lo Nigro, R., Rimini, E., and Raineri, V. (2011). Mapping the density of scattering centers limiting the electron mean free path in graphene, *Nano Lett.*, **11**, pp. 4612–4618.
19. Ristein, J., Mammadov, S., and Seyller, T. (2012). Origin of doping in quasi-free-standing graphene on silicon carbide, *Phys. Rev. Lett.*, **108**, 246104.
20. Tedesco, J. L., VanMil, B. L., Myers-Ward, R. L., McCrate, J. M., Kitt, S. A., Campbell, P. M., Jernigan, G. G., Culbertson, J. C., Eddy, Jr., C. R., and Gaskill, D. K. (2009). Hall effect mobility of epitaxial graphene grown on silicon carbide, *Appl. Phys. Lett.*, **95**, 122102.
21. Hiebel, F., Mallet, P., Varchon, F., Magaud, L., and Veuillen, J.-Y. (2008). Graphene-substrate interaction on 6H-SiC(000-1): A scanning tunneling microscopy study, *Phys. Rev. B*, **78**, 153412.
22. Borysiuk, J., Bozek, R., Grodecki, K., Wysmołek, A., Strupinski, W., StePniewski, R., and Baranowski, J. M. (2010). Transmission electron microscopy investigations of epitaxial graphene on C-terminated 4H–SiC, *J. Appl. Phys.*, **108**, 013518.
23. Colby, R., Bolen, M. L., Capano, M. A., and Stach, E. A. (2011). Amorphous interface layer in thin graphite films grown on the carbon face of SiC, *Appl. Phys. Lett.*, **99**, 101904.
24. Weng, X., Robinson, J. A., Trumbull, K., Cavalero, R., Fanton, M. A., and Snyder, D. (2012). Epitaxial graphene on SiC(000-1): Stacking order and interfacial structure, *Appl. Phys. Lett.*, **100**, 031904.
25. Camara, N., Rius, G., Huntzinger, J.-R., Tiberj, A., Magaud, L., Mestres, N., Godignon, P., and Camassel, J. (2008). Early stage formation of graphene on the C face of 6H-SiC, *Appl. Phys. Lett.*, **93**, 263102.
26. Bouhafs, C., Darakchieva, V., Persson, I. L., Tiberj, A., Persson, P. O. A., Paillet, M., Zahab, A. A., Landois, P., Juillaguet, S., Schoche, S., Schubert, M., and Yakimova, R. (2015). Structural properties and dielectric

function of graphene grown by high-temperature sublimation on 4H-SiC(000-1), *J. Appl. Phys.*, **117**, 085701.

27. Nicotra, G., Deretzis, I., Scuderi, M., Spinella, C., Longo, P., Yakimova, R., Giannazzo, F., and La Magna, A. (2015). Interface disorder probed at the atomic scale for graphene grown on the C-face of SiC, *Phys. Rev. B*, **91**, 155411.
28. Vecchio, C., Sonde, S., Bongiorno, C., Rambach, M., Yakimova, R., Rimini, E., Raineri, V., and Giannazzo, F. (2011). Nanoscale structural characterization of epitaxial graphene grown on off-axis 4H-SiC (0001), *Nanoscale Res. Lett.*, **6**, 269.
29. Giannazzo, F., Deretzis, I., La Magna, A., Roccaforte, F., and Yakimova, R. (2012). Electronic transport at monolayer-bilayer junctions in epitaxial graphene on SiC, *Phys. Rev. B*, **86**, 235422.
30. Ferrari, A. C., Meyer, J. C., Scardaci, V., Casiraghi, C., Lazzeri, M., Mauri, F., Piscanec, S., Jiang, D., Novoselov, K. S., Roth, S., and Geim, A. K. (2006). Raman spectrum of graphene and graphene layers, *Phys. Rev. Lett.*, **97**, 187401.
31. Lee, D., Riedl, C., Krauss, B., von Klitzing, K., Starke, U., and Smet, J. H. (2008). Raman spectra of epitaxial graphene on SiC and of epitaxial graphene transferred to SiO_2, *Nano Lett.*, **8**, pp. 4320–4325.
32. Krivanek, O. L., Chisholm, M. F., Nicolosi, V., Pennycook, T. J., Corbin, G. J., Dellby, N., Murfitt, M. F., Own, C. S., Szilagy, Z. S., Oxley, M. P., Pantelides, S. T., and Pennycook, S. J. (2010). Atom-by-atom structural and chemical analysis by annular dark-field electron microscopy, *Nature*, **464**, pp. 571–574.
33. Yakes, M. K., Gunlycke, D., Tedesco, J. L., Campbell, P. M., Myers-Ward, R. L., Jr. Eddy, C. R., Gaskill, D. K., Sheehan, P. E., and Laracuente, A. R. (2010). Conductance anisotropy in epitaxial graphene sheets generated by substrate interactions, *Nano Lett.*, **10**, pp. 1559–1562.
34. Giannazzo, F., Deretzis, I., Nicotra, G., Fisichella, G., Spinella, C., Roccaforte, F., and La Magna, A. (2014). Electronic properties of epitaxial graphene residing on SiC facets probed by conductive atomic force microscopy, *Appl. Surf. Sci.*, **291**, pp. 53–57.
35. Odaka, S., Miyazaki, H., Li, S.-L., Kanda, A., Morita, K., Tanaka, S., Miyata, Y., Kataura, H., Tsukagoshi, K., and Aoyagi, Y. (2010). Anisotropic transport in graphene on SiC substrate with periodic nanofacets, *Appl. Phys. Lett.*, **96**, 062111.
36. Giannazzo, F., Deretzis, I., Nicotra, G., Fisichella, G., Ramasse, Q. M., Spinella, C., Roccaforte, F., and La Magna, A. (2014). High resolution

study of structural and electronic properties of epitaxial graphene grown on off-axis 4H-SiC (0001), *J. Cryst. Growth*, **393**, pp. 150–155.

37. Jouault, B., Jabakhanji, B., Camara, N., Desrat, W., Tiberj, A., Huntzinger, J.-R., Consejo, C., Caboni, A., Godignon, P., Kopelevich, Y., and Camassel, J. (2010). Probing the electrical anisotropy of multilayer graphene on the Si-face of 6H-SiC, *Phys. Rev. B*, **82**, 085438.
38. Schumann, T., Friedland, K.-J., Oliveira, M. H., Tahraoui, A., Lopes, J. M. J., and Riechert, H. (2012). Anisotropic quantum Hall effect in epitaxial graphene on stepped SiC surface, *Phys. Rev. B*, **85**, 235402.
39. Ji, S.-H., Hannon, J. B., Tromp, R. M., Perebeinos, V., Tersoff, J., and Ross, F. M. (2011). Atomic-scale transport in epitaxial graphene, *Nat. Mater.*, **11**, pp. 114–119.
40. Nagase, M., Hibino, H., Kageshima, H., and Yamaguchi, H. (2009). Local conductance measurements of double-layer graphene on SiC substrate, *Nanotechnology*, **20**, 445704.
41. Giannazzo, F., Deretzis, I., La Magna, A., Nicotra, G., Spinella, C., Fisichella, G., Fiorenza, P., Yakimova, R., and Roccaforte, F. (2015). Origin of the current transport anisotropy in epitaxial graphene grown on vicinal 4H-SiC (0001) surfaces, *Mater. Sci. Forum*, **806**, pp. 103–107.
42. Hicks, J., Tejeda, A., Taleb-Ibrahimi, A., Nevius, M. S., Wang, F., Shepperd, K., Palmer, J., Bertran, F., Le Fèvre, P., Kunc, J., de Heer, W. A., Berger, C., and Conrad, E. H. (2013). A wide-bandgap metal–semiconductor–metal nanostructure made entirely from graphene, *Nat. Phys.*, **9**, pp. 49–54.
43. Baringhaus, J., Tegenkamp, C., Edler, F., Ruan, M., Conrad, E., Berger, C., and de Heer, W. A. (2014). Exceptional ballistic transport in epitaxial graphene nanoribbons, *Nature*, **506**, pp. 349–354.
44. Giannazzo, F., Sonde, S., Raineri, V., and Rimini, E. (2009). Irradiation damage in graphene on SiO_2 probed by local mobility measurements, *Appl. Phys. Lett.*, **95**, 263109.
45. Giannazzo, F., Musumeci, P., Calcagno, L., Makhtari, A., and Raineri, V. (2001). Carrier concentration profiles in 6H-SiC by scanning capacitance microscopy, *Mater. Sci. Semicond. Process.*, **4**, pp. 195–199.
46. Giannazzo, F., Goghero, D., Raineri, V., Mirabella, S., and Priolo, F. (2003). Scanning capacitance microscopy on ultranarrow doping profiles in Si, *Appl. Phys. Lett.*, **83**, 2659.
47. Giannazzo, F., Sonde, S., Raineri, V., and Rimini, E. (2009). Screening length and quantum capacitance in graphene by scanning probe microscopy, *Nano Lett.*, **9**, pp. 23–29.

Chapter 5

Theory of Graphene Growth on SiC Substrate

Hiroyuki Kageshima
Interdisciplinary Graduate School of Science and Engineering, Shimane University, 1060 Nishi-Kawatsucho, Matsue, Shimane 690-8504, Japan
kageshima@ecs.shimane-u.ac.jp

5.1 Introduction

By definition, graphene is a one atom–thick, two-dimensional, crystalline material, consisting entirely of carbon atoms. Electronically, pristine graphene is a zero-gap semiconductor [1–6]. Because of its unique properties such as very high mobility at room temperature, it is expected to play an important role as the next-generation electronic device material [7–20].

For instance, high-quality graphene film can be obtained by thermal treatment of silicon carbide SiC [21–27]. When an SiC substrate is annealed at temperatures higher than about 1000°C, graphene can be deposited on the substrate surface, as a result of the thermal crystal decomposition of the SiC substrate. The melting

Epitaxial Graphene on Silicon Carbide: Modeling, Devices, and Applications
Edited by Gemma Rius and Philippe Godignon

ISBN 978-981-4774-20-8 (Hardcover), 978-1-315-18614-6 (eBook)
www.panstanford.com

temperature of the silicon (Si) is 1420°C; i.e., significantly lower than the melting temperature of SiC (2730°C), as well as the sublimation temperature of graphite, which for solid carbon is 3642°C. Applied to SiC, Si atoms can be selectively sublimated from the surface of SiC, and the excess C atoms remaining on the surface may lead to the formation of graphene materials. The most thermodynamically stable C solid at standard conditions is the graphite material, namely a three-dimensional crystal consisting of sequentially stacked graphene sheets [28].

Considering the process of graphene formation on SiC at the atomic scale is a complex problem. Roughly, on the one hand, C atoms aggregate on the surface to form graphene. This process is similar to the epitaxial growth process commonly accepted [29]. On the other hand and simultaneously, Si atoms sublimate from the surface. This process can be thought as an etching process (of the SiC crystal surface). Summarizing, the mechanism for the formation of graphene by high-temperature treatment of SiC relies on two completely different atomic processes, which coexist and combine on the same sample surface.

It is widely accepted that atomically thick films with high crystal quality can be formed by the so-called Frank–van der Merwe growth, known as the general epitaxial growth process [29]. The Frank–van der Merwe growth considers that adatoms attach preferentially to surface step edge sites and form an atomically smooth layer. According to that, two-dimensional layered materials grow sequentially, layer by layer. It is also widely accepted that a smooth surface can be obtained via the step backflow of the surface based on conventional etching processes [30]. In this case, atoms along the step edges sublimate so that the steps move backward. Therefore, it is expected that the Frank–van der Merwe growth and the conventional atomic etching process coexist to lead to the graphene growth by the high-temperature treatment of SiC.

Here, our understanding of the graphene growth mechanism on SiC obtained from the first-principles study is compiled [31–38]. In this chapter, we focus mainly on the SiC(0001) Si-face, because most experimental studies have been carried out on this SiC surface termination.

5.2 Fundamental Aspects

In this section, first, some fundamental aspects are briefly summarized before discussing the growth mechanism itself.

5.2.1 Dependence of the Stacking Ordering of Graphene Layers on the Termination of SiC

Experimentally, the formation of graphene crystal clearly depends on the surface termination of the SiC substrate. The atomic structure and the surface terminations of SiC crystal are shown in Fig. 5.1. Most commonly, for (000-1) C-face, rather thick multilayer graphene is formed. The obtained graphene is turbostratic: The graphene sheets are randomly stacked upon the substrate, and the topmost graphene sheet typically exhibits monolayer graphene electronic behavior [21, 39–41]. Differently, for the (0001) Si-face, monolayer or a few-layer graphene is typically formed [23, 25, 41–54]. In this case, the grown graphene is epitaxial, in other words, the positions of C atoms forming the graphene layer are intimately related to the SiC substrate. Recent studies also report that graphene on the (11-20) a-face is of epitaxial type [55–60].

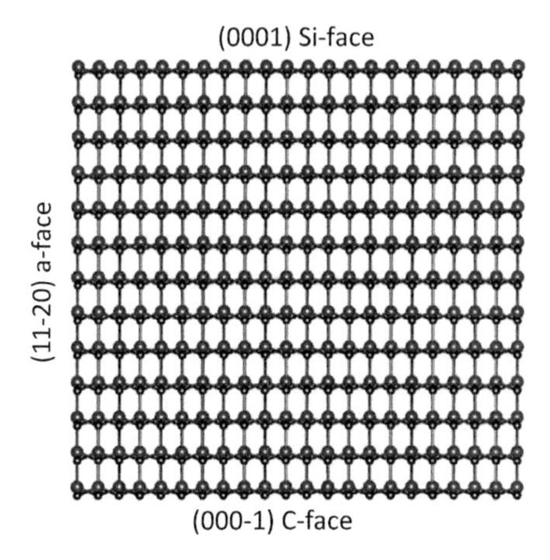

Figure 5.1 Atomic structure and surface terminations of 6H-SiC crystal. Large and small spheres indicate Si and C atoms, respectively.

5.2.2 Dependence of the Atomic Ordering of Graphene on the Graphene–SiC Interface

The interfacial structure between the graphene and the SiC substrate is also dependent on the starting surface termination of the SiC substrate. The main difference is that on the Si-face, a buffer layer is formed on the substrate before graphene layer deposition [61]. It has been reported that post-hydrogen annealing treatment converts the buffer layer into an (additional) graphene layer, which is called a quasi-freestanding layer [62]. This converted additional layer acts as a monolayer graphene in the carrier transport measurements [63]. Therefore, the buffer layer is like a graphene sheet, but tightly bounded with the substrate surface. The post-hydrogen annealing treatment terminates the chemical bonds of the substrate surface with hydrogen atoms, so that the buffer layer leaves from the supporting SiC substrate and floats on the substrate surface. This buffer-layer graphene is called the zeroth graphene layer. Such a buffer layer, however, is not observed for the C-face graphene growth [64–68]. The buffer layer is neither observed for the a-face [60].

The buffer layer structure is also related with the doping of the deposited graphene. The graphene on the buffer layer of the Si-face is negatively charged, and the carrier transport measurement observes conduction electrons as the majority carrier at the nearly zero gate bias [69–71]. The quasi-freestanding layer is slightly positively charged. Instead, the graphene on the C-face is almost neutral.

5.2.3 Dependence of the Atomic Ordering of Graphene on the Temperature and Si Pressure

The general sublimation process is a stepped sequence with temperature increase. In the case of the Si-face, the SiC substrate is initially reconstructed as a $\sqrt{3}\times\sqrt{3}$ structure generally [26]. On increasing the temperature, the surface is converted into $6\sqrt{3}\times6\sqrt{3}$ and then into a couple of layers of graphene. The $6\sqrt{3}\times6\sqrt{3}$ reconstructed surface corresponds to the surface covered with the zeroth graphene layer. The transition temperatures of the surface structure phases change with the Si pressure (Fig. 5.2) [26, 72–79]. An increase in the Si pressure raises the transition temperatures.

Some experiments also report finger-like structure surfaces, in which surface steps of the substrate wander considerably [73–75, 77].

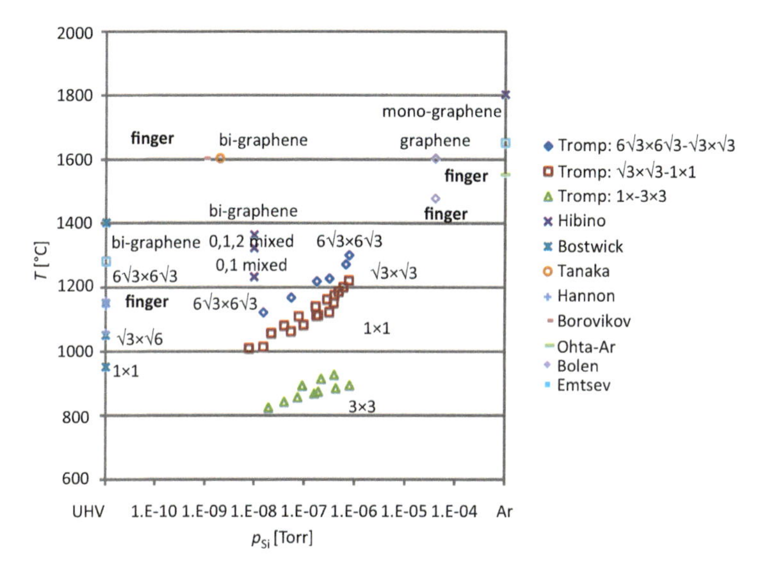

Figure 5.2 Summary of experimentally reported conditions to form various surface structure of SiC(0001), Si-face, with the sublimation process. The experiments are reported in Tromp [76], Hibino [26], Bostwick [72], Tanaka [78], Hannon [73], Borovikov [75], Ohta-Ar [77], Bolen [74], and Emtsev [79].

5.2.4 Dependence of the Atomic Ordering of Graphene on Atmosphere

Experiments also show the dependence of graphene crystallinity on atmosphere at the sublimation process [26]. Under ultrahigh vacuum (UHV) conditions, the sublimation process can form bilayer graphene on the surface. The formed bilayer graphene may cover wide areas even over many SiC substrate steps [45, 46]. On the other hand, in the high-pressure argon (Ar) atmosphere, the sublimation process tends to create a monolayer or just a zero-layer (buffer-layer) graphene [79].

Detailed experimental studies, however, reveal a distinctive crystal quality between these two conditions [71]. The graphene formed in the UHV condition has relatively smaller domains than that synthesized under Ar condition. The difference is attributed to the conditions of the growth at the atomic scale. In the UHV condition, the initially atomically flat $\sqrt{3}\times\sqrt{3}$ surface becomes rough when the $6\sqrt{3}\times6\sqrt{3}$ surface reconstruction is formed. After monolayer and bilayer graphene islands are deposited, from the step edges, the whole surface is covered with a bilayer graphene and returns to be atomically flat. In the case of Ar condition, such roughening of the surface is not observed. The surface is kept smooth during the graphene formation. It is also known that oxygen in the annealing atmosphere induces carbon nanotube rather than graphene [80, 81]. The ambient is thus important for graphene growth.

5.2.5 Relation between SiC Crystal Etching and Graphene Growth Rate

As the last remark of this section, the relation between etching and growth speed is considered. In the zeroth graphene layer structure, namely the $6\sqrt{3}\times6\sqrt{3}$ structure, of the Si-face, the surface of the resulting graphene sheet is smooth, so is the substrate. A $6\sqrt{3}\times6\sqrt{3}$ structure relates to the unit cell of the SiC Si-face, meaning that the structure is $6\sqrt{3}\times6\sqrt{3}$ times the primitive cell of the SiC Si-face. The graphene sheet has a unit cell with 13×13 graphene primitive cells for each unit cell of the $6\sqrt{3}\times6\sqrt{3}$ structure [61, 82]. One primitive cell of the graphene consists of two C atoms, while one bilayer of one primitive cell of the SiC Si-face consists of one Si and one C atoms. This means that one graphene layer is formed of $13 \times 13 \times 2 = 338$ C atoms, while one bilayer of SiC Si-face includes $6\sqrt{3} \times 6\sqrt{3} = 108$ C atoms per "unit cell" area. Therefore, to form a single monolayer graphene sheet, it is necessary to etch at least $338/108 = 3.13$ bilayers of the Si-face. This means that the SiC etching rate is at least three times faster than the graphene growth rate.

5.3 Sublimation from Terraces and Immobile Carbon

The effect of the Si sublimation is examined next, as the first step for establishing a theory on the graphene growth mechanism for the Si-face [33]. The surface is assumed to be completely (atomically) flat, and C atoms are considered to be immobile on the surface. In other words, the surface would have no step, and excess C atoms, resulting from the Si atoms sublimation, remain and aggregate at the same atomic location where the Si atoms sublimate.

A first-principles approximation has been applied to a 3×3 supercell. An atomically flat surface is considered the initial surface. After one Si atom is sublimated from the surface, the substrate atomic structure is optimized according to the most energetically stable structure. This sequence is iterated until a bilayer graphene film is formed on the surface.

The results show that excess C atoms and their aggregates are kept bound to the surface until they arrange into a monolayer graphene sheet. Once a monolayer graphene sheet is formed, the sheet is released from the surface and become a quasi-freestanding sheet. After the monolayer graphene sheet is formed, newly released C atoms aggregate on the SiC surface underneath the quasi-freestanding graphene sheet. Nucleation of that C atoms is similar to the initial, which formed without a covering monolayer graphene sheet. As a result, another monolayer graphene sheet is formed, so the surface is covered with a bilayer graphene.

However, the actual formation of bilayer graphene is not as simple. Since the etching speed is about three times faster than the graphene growth rate, as described earlier, the SiC surface is, effectively, substantiantially deconstructed during the bilayer deposition sequence. For instance, the released excess C atoms can aggregate with both Si atoms as well as with other excess C atoms. Additionally, in some cases, excess C atoms can also become trapped in the vacancy site of the sublimated Si atoms. As another possibility, an Si atom can get trapped within clustered C atoms. Energetically, this means that graphene-formation energy largely oscillates with an increase in the sublimation rate Si atoms. Under those conditions, it is expected that the graphene deposited is of low crystal quality. Conversely, in order to obtain high-quality graphene, it has been suggested that C nucleation, to form graphene flms, should proceed in lattice sites different from those in which the atomic Si sublimation takes place.

5.4 Carbon Growth on Terrace

Based on the previous premise for obtaining nucleation high-quality graphene, we consider only the process of crystal growth on the terrace of the SiC Si-face [31, 32]. Therefore, we do not consider how Si atoms sublimate from the surface here. Instead, we just assume that Si atoms sublimate somewhere by which C atoms are provided on a specific terrace.

Again, first-principles calculations are performed on a 3×3 supercell. We formulate an atomically flat Si-face as the initial surface; we distribute C atoms on the surface according to a series of configurations and optimize them atomically as graphitic structures to search the most stable atomic configuration(s) for the proposed coverage.

We have found that adsorbed C atoms prefer to bind covalently with the topmost Si atoms of the surface. As C atoms coverage increases, C monomers can start forming, dimerstrimer, etc., sequentially. As a result an atomic C chain is formed, to which additional C atoms can be attached. This is how two-dimensional, sheet-like, C network is formed—a monolayer graphene film. Since the monolayer graphene still has covalent bonds with the topmost Si atoms on the substrate, this one atom-thick layer will not exhibit the Dirac cone in its band dispersion structure, as expected for freestanding graphene [82–84]. Instead, its electronic structure corresponds to the buffer layer structure, the $6\sqrt{3}\times6\sqrt{3}$ structure, or zeroth graphene structure.

The topmost Si atoms on the substrate thus act as a constraint for C atoms to aggregate and to evolve into a monolayer graphene. Relevantly, since only C atoms aggregate and any Si atoms do not bind to other Si atoms, this promotes the formation of monolayer graphene.

In our calculations, the formation energies of the monolayer and the bilayer graphene are much more stable than the structure with the other C aggregates such as a dimer. Therefore, the C aggregates tend to grow as graphene flakes, which become larger and larger by attaching new C atoms to the edge. Aware of the limitations of our calculation methods, our results suggest that an additional new graphene sheet would always grow at the interface between an existing graphene sheet and the SiC substrate (Fig. 5.3). Consequently, the grown graphene sheets would be lifted upward, away from the substrate, when a new graphene sheet is growing. Therefore, the existing graphene sheet would remain as the topmost even when few-layer/multilayer graphene is formed. Because the growing graphene film always grows directly on the substrate, as a kind of buffer layer, even bilayer/few-layer graphene will correspond to an epitaxial process.

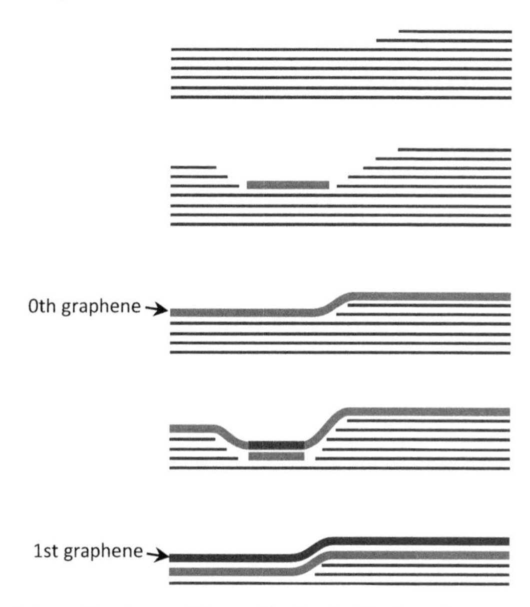

Figure 5.3 Schematic views of theoretically clarified graphene growth process on SiC(0001) Si-face.

Importantly, our interfacial growth model is consistent with Si isotope experiments [85]. This model is also supported by experimental observations, where the deposition rate for the first layer is faster than subsequent ones on the Si-face of SiC substrates [78]. This is due to the fact that as the surface is covered with some/various graphene sheets, the sublimated Si from the topmost of the Si-face should diffuse but not escape from the surface as the covering graphene sheets hinder their sublimation. The thicker and more complete the graphene film, the lower the graphene deposition rate.

5.5 Step and Silicon Sublimation

In the previous sections, we neglected steps, which are generally very important for crystal growth considerations. In this section, we focus on the role of the SiC steps in the Si sublimation process [35, 36].

The steps are thought to be the predominant sites for Si sublimation as well as preferential C nucleation sites (Fig. 5.4). To investigate the role of the steps, we prepare a trench-based model consisting of a lower and a higher terrace as the initial surface structure. Because, locally, the edges of the two terraces can be straight single-atom steps, we use this as the basic element of our model for the steps. One of the trench models consists of a $4\sqrt{3}\times\sqrt{3}$ supercell and [11–20] steps. The other is a model with a $2\times4\sqrt{3}$ supercell, a Si[1–100] step, and a C[1–100] step. We remove one Si atom from various positions of the topmost surface to find which position is preferable for the Si sublimation. Then, first-principles calculations are repeated.

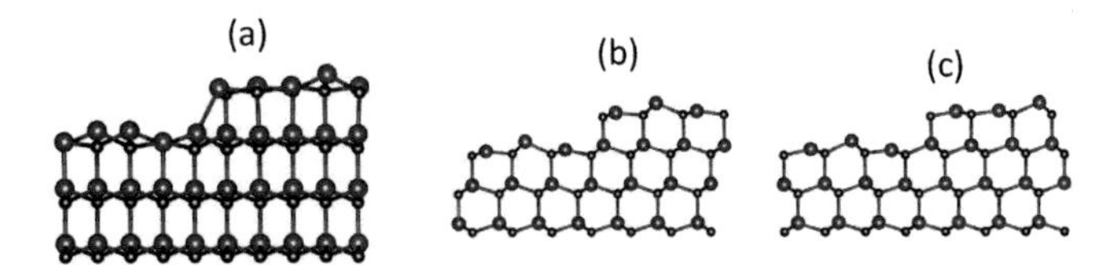

Figure 5.4 Side views of the atomic structures of the steps on the SiC(0001) Si-face. (a) [11–20] step, (b) Si-[1-100] step, and (c) C-[1-100] step. Large and small spheres indicate Si and C atoms, respectively.

5.5.1 On Pristine SiC Surface

First we calculated the pristine surface case. The calculations indicate that Si atoms do not sublimate preferentially from the step edge, for any step orientations, in contrast to the intuitive considerations. Instead, an Si atom preferably sublimates from the center of the terrace. This fact implies that the substrate surface tends to become rough due to the atomic Si sublimation process, unless some additional mechanism takes part.

Because the step edge consists of both C atoms and Si atoms, C atoms are thought to stabilize the step edge. Therefore, for the calculations, we remove one C atom at the step edge first, then examine the required energy to sublimate one Si atom from the step edge. The result of the calculation indicates that the energy for the Si sublimation is negative in this case. This means that an Si atom can spontaneously sublimate from the step edge once a C atom is desorbed from the step edge.

The preferential atomic C desorption from the step edge, as compared with the atomic Si desorption from the center of the terrace, can be controlled by the Si partial pressure. If the Si pressure is high, desorption of Si atoms is not promoted.

We consider that the surface is in equilibrium with Si gas, isolated graphene, and bulk SiC crystal. Here we neglect the presence of gas other than Si gas [86]. From statistical physics, the chemical potential $d\mu_{\mathrm{C}}$ of the C atoms on the surface can be written as

$$d\mu_{\mathrm{C}} = (E_{\mathrm{SiC}} + \mu_{\mathrm{SiC}}^{\mathrm{phonon}}) - (E_{\mathrm{atSi}} + \mu_{\mathrm{atSi}}^{\mathrm{gas}}) - E_{\mathrm{grpn}} \tag{5.1}$$

where E_{SiC} is the total energy of bulk SiC crystal per SiC unit cell, E_{atSi} is the total energy of one isolated Si atom, E_{grpn} is the total energy of graphene per single C atom, $\mu_{\mathrm{SiC}}^{\mathrm{phonon}}$ is the chemical potential of one phonon of bulk SiC crystal, and $\mu_{\mathrm{atSi}}^{\mathrm{gas}}$ is the chemical potential of the Si gas. Similarly, the chemical potential $d\mu_{\mathrm{Si}}$ of the Si atoms on the surface can be written as

$$d\mu_{\mathrm{Si}} = E_{\mathrm{atSi}} + \mu_{\mathrm{atSi}}^{\mathrm{gas}} - (E_{\mathrm{SiC}} - E_{\mathrm{grpn}}) \tag{5.2}$$

By comparing Eqs. (5.1) and (5.2), $d\mu_{\mathrm{C}}$ can be rewritten as

$$d\mu_{\mathrm{C}} = -d\mu_{\mathrm{Si}} + \mu_{\mathrm{SiC}}^{\mathrm{phonon}} \tag{5.3}$$

Therefore, $d\mu_{\mathrm{C}}$ can be controlled by $d\mu_{\mathrm{Si}}$, which, in turn, is simply controlled by Si gas $\mu_{\mathrm{atSi}}^{\mathrm{gas}}$, which can be written as

$$\mu_{\mathrm{atSi}}^{\mathrm{gas}} = -k_{\mathrm{B}}T \log\left[g \frac{k_{\mathrm{B}}T}{p_{\mathrm{Si}}} \left(\frac{2\pi m k_{\mathrm{B}} T}{h^2} \right)^{3/2} \right] \tag{5.4}$$

where T is the temperature, p_{Si} is the pressure of the Si gas, m is the mass of an Si atom, k_{B} is the Boltzmann constant, h is the Planck constant, and $g = 3$ is the degeneracy of the electronic states of an Si atom [87]. Employing the Debye model, $\mu_{\mathrm{SiC}}^{\mathrm{phonon}}$ can also be written as

$$\mu_{\text{SiC}}^{\text{phonon}} = \frac{1}{3N}\omega_{\text{D}} k_{\text{B}} T \log\left[2\sinh\frac{\hbar\omega_{\text{D}}}{2k_{\text{B}}T}\right] D(\omega_{\text{D}}) \tag{5.5}$$

where N is the number of Si–C atom pairs in the bulk SiC crystal, ω_{D} is the Debye frequency [88], $D(\omega_{\text{D}})$ is the density of states of the phonon of the bulk SiC crystal at the Debye frequency, and $\hbar = h/2\pi$ is the reduced Planck constant.

Then, in order to destabilize the most stable C-[1–100] step, it is required that

$$d\mu_{\text{Si}} - d\mu_{\text{C}} < 0.41[\text{eV}]. \tag{5.6}$$

Based on these equations, a critical temperature is determined as 1120°C for an Si pressure of 10^{-8} Torr. When the temperature is lower than this critical temperature, the Si atoms would sublimate from the step edges. For an Si pressure of 0.02 Torr, the critical temperature would be 1800°C. When the Si partial pressure is higher than this critical value, Si atoms tend to sublimate from the step edge. As a result, sublimation from the step edges leads to the backflow of the steps, which materializes as a reduction of the step density and promotes surface smoothness. On the other hand, when the Si pressure is lower, or the temperature is higher, than the corresponding critical value, the Si atoms tend to sublimate from terrace surface sites. Preferential sublimation from the terrace implies an increase in the step density and the surface roughening. A complete theoretical phase diagram of the surface morphology during the Si sublimation is shown in Fig. 5.5. Experimental UHV condition, with the Si pressure lower than 10^{-8} Torr, is certainly within the surface roughening phase. On the other hand, experimental conditions of high-pressure Ar flow at a temperature of 1800°C correspond to the surface smoothening phase, assuming that the presence of Ar efficiently enhances the Si partial pressure.

As the last remark, Fig. 5.5 also shows that the theoretical critical line almost coincides with the transition between the $\sqrt{3}\times\sqrt{3}$ and the $6\sqrt{3}\times6\sqrt{3}$ (zeroth graphene) surface phases found experimentally. In our theoretical model, the step stability is overestimated because only the case of a straight step is considered. Therefore, more precise theoretical calculation of the critical line will be at a higher temperature.

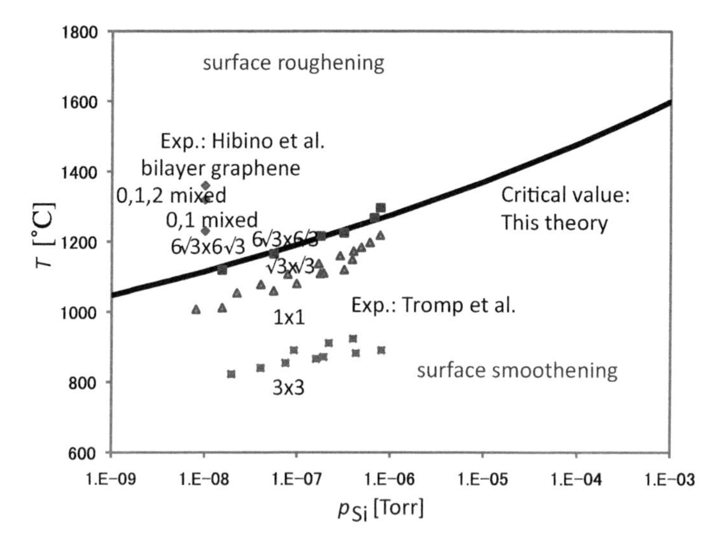

Figure 5.5 Theoretical phase diagram of the surface morphology for pristine SiC(0001) Si-face during the Si sublimation. Comparative experimental results from Hibino et al. [26] and Tromp et al. [76] are also shown.

5.5.2 Post-Zeroth Graphene Layer Formation

We have discussed atomic Si sublimation in the process of the first graphene layer formation on a pristine SiC surface, namely the zeroth graphene layer growth process. Next we study the Si sublimation after the zeroth graphene layer has been formed. For that purpose, we consider an initial surface completely covered with a monolayer of graphene sheet. The stable energies of the [11–20], Si-[1–100], and C-[1–100] steps, again, are calculated by removing one Si atom located at various surface sites. The results indicate that the step is unstable, whatever the step orientation. Si desorption from the step edge requires less energy than that located on the terrace. This means that the steps tend to flow backward once the whole surface is covered with the zeroth graphene sheet. The instability of the steps can be attributed to the modification of the step atomic configuration of the Si atoms as covering graphene sheet is present.

This theoretical finding is also consistent with the experimental results. In the high-pressure Ar atmosphere, experiments show that the surface remains flat even after the zeroth graphene is formed.

Instead, in a UHV condition, experiments show that the surface would recover the smoothness during the formation of the bilayer graphene, even when severe surface roughening during the zeroth graphene growth occurred.

5.6 Role of Steps and Carbon Nucleation

5.6.1 Initial Stage of Carbon Nucleation and the Role of Steps

First we consider the initial stage of the C nucleation. We employ the trench models as in the previous section. We put one or two C atoms on the various sites of the ideal surfaces and calculate the optimal atomic arrangement and their corresponding formation energies.

In the case of the pristine surface, the first C atom preferentially adsorbs within/on the terrace for the [11–20] step, while it preferentially adsorbs at the lower step edge for the Si-[1–100] and C-[1–100] steps. However, even for the [11–20] step, once the second C atom is adsorbed, the two C atoms preferentially aggregate at the lower step edge rather than to remain on the terrace. This indicates that C atoms start to aggregate at the step edge when a certain amount of atomic C is present and provided that can sufficiently diffuse on the surface.

The results of calculations show basically the same situation after the whole surface is covered with a monolayer graphene sheet. In this case, the first C atom can adsorb at the lower step edge even for the [11–20] step, as well as for the other orientations.

In other words, the lower step edge is a good site for C nucleation. This behavior is consistent with the experimental results. In experiments under UHV, the C growth starting at the lower step edge has been observed [26, 73]. It is also found that the second and the third graphene layers grow simultaneously when the step is high enough, as a result of surface roughening.

Therefore, the control upon the initial stage is the most important or critical condition to obtain high-quality graphene. The formed zeroth graphene would be polycrystalline if deposited upon surface roughening conditions of Si sublimation. Consequently, as the first graphene layer remains, it becomes an important factor for electronic device applications such as the top-gate field-effect transistors, even if thick graphene films are formed.

5.6.2 Effect of Epitaxy at the Step

In order to consider the subsequent graphene growth process, we study the formation of graphene islands on the terrace and at the lower step edge. It has been theoretically revealed that the C nucleates become a C atomic ring in the freestanding case [89, 90]. Therefore, it must be important to consider lattice mismatch for the graphene growth.

First we examine the C island growth on the clean surface [38]. The calculations indicate that the C aggregate surely forms graphene-like network. This is thought to come from the supporting nature of the SiC surface. The dangling bonds of the surface Si trap the adsorbed C atoms and induce the formation of the C sp^2 network rather than that of the C *sp* network. However, the C aggregate preferentially contains defects such as five-membered rings or seven-membered rings in addition to the six-membered rings held in the perfect graphene. This suggests that the supporting nature of the SiC terrace is too strong to obtain perfect graphene. Because the Si dangling bond network of the SiC surface does not completely match the graphene network, the formed graphene island contains defects to keep the strong Si–C bonds.

The step is thought to assist the formation of graphene without defects by a template effect [91]. When the graphene island grows at the lower step edge, the calculation reveals that it is tightly bound at the lower step edge and slightly floats from the terrace. This makes the graphene island free from the defects. Because the C atoms of the graphene island are more loosely bound by the surface Si atoms of SiC, the graphene island can easily have a perfect network only with the six-membered rings.

5.7 Further Discussions

Graphene growth mechanism has already been discussed, but the discussions are limited to a rather primitive stage. Some further discussions are as follows.

5.7.1 Step Wandering

Experiments often observe step wandering during graphene growth [73–75, 77]. Some experiments observe finger structures with the highly wandering steps. The reason is thought to be the instability of the step. As discussed earlier, the step edge becomes the C nucleation site as well as the Si sublimation site. The C nucleation could modify the step flow, while the Si sublimation just induces it [92].

Remarkably, exiting theoretical reports successfully explain step wandering [5]. It is based on the atomic Si diffusion toward the terrace as well as the atomic Si mobility along the step. In addition, it takes into account higher Si sublimation rate at concave-like step regions. Essentially, this model shows that the growth mechanism changes from the no-growth mode into an unstable-step mode, and then to a stable-step mode, as the temperature increases. The theory also shows how the pressure relates to the mode diagram. However, this theory does not explicitly consider the role of the formed graphene islands on the surface.

In brief, it has not been clarified how the formed graphene islands correlate with the steps. Some experiments [73] report that the graphene islands are initially formed at the lower step edge, but that their positions asymptotically change into the upper step edge. While the graphene island stays at the same place as that initially formed, the backflow of the steps is thought to change its relative position with the step. Further studies are necessary to elucidate step wandering to enable a correlation with the formed graphene islands.

5.7.2 Silicon Sublimation Path

As graphene layers form on the substrate, eventually the whole surface becomes fully covered with a monolayer graphene. Then, it is not clear how the Si atoms can continue to sublimate from the SiC surface.

Effectively, some experiments report that the graphene growth rate is reduced due to the covering graphene sheet [80]. They discuss the reduced growth rate and explain it by the Si diffusion through the growing graphene sheet. However, on a zero-structural-defect graphene sheet, the occurrence of Si atoms sublimation is difficult to be accepted. A theoretical study reports that even graphene structure with seven-membered rings also shows a very high-energy barrier, over 4 eV, for an Si atom to be able to escape through [93]. Instead, one possibility is that Si atoms diffuse through larger holes of the graphene sheet, but that the holes are repaired and completely swept away after the Si sublimation process ceases. Actually, it is found experimentally that Si reduces graphene and makes holes [94, 95]. Yet further studies are necessary.

5.7.3 Orientation Dependence

As shown in Section 5.2, crystallinity of the formed graphene strongly depends on the surface orientation of the SiC substrate. However, the reason is not exactly clarified, yet.

For what it is known, the buffer layer is thought to relate with the crystallinity. The buffer layer is formed on the (0001) Si-face, where the epitaxial graphene is subsequently formed; differently, the buffer layer is not formed on the (000-1) C-face, where turbostratic graphene is formed. It is also reported that graphene deposition on the SiC (11–20) does not follow this rule [60]. The formed graphene is, indeed, epitaxial, but the buffer layer has not been observed. However, a recent theoretical study reveals that the buffer layer is formed on the surface at the initial stage of graphene growth [37]. Therefore, it seems that the buffer layer formation can be required and often relates with graphene crystallinity.

However, the fundamental reason why the buffer layer is formed on the (0001) Si-face, but not on the (000-1) C-face, is not clear. The control upon polar faces of bulk SiC might be the key to solve this question [96].

5.7.4 Kinetics and Dynamics

In the previous sections, the studies are limited to energetics and thermal equilibrium cases. To deepen the comprehension, the kinetics of the Si sublimation and the C nucleation should be clarified. Dynamic processes of the Si atoms and the C atoms should be clarified, too.

Theoretical studies using kinetic Monte Carlo simulations and rate equations reveal that two different growth regimes appear depending on variables such as total coverage, vicinal angle, and model parameters [97, 98]. For instance, one growth regime is dominated by the coalescence of graphene strips, while the other is ruled by a process for graphene "climbing" over step edges. On those papers, the results are supported with experimental results. These investigations require that diffusion, sublimation, and nucleation processes of Si and C atoms on the surface, as well as the formation mechanisms for graphene islands and strips are precisely considered.

5.8 Summary

In this chapter, the graphene growth mechanism has been theoretically discussed, mainly for graphene deposition on the SiC(0001) Si-face. The Si sublimation occurs preferentially from the step edge, but strongly depending on the sample condition. The graphene sheet grows from the step edge and epitaxially on the SiC substrate surface even if the surface is covered with graphene sheets. The models, however, are limited to rather primitive stages and simplified scenarios. Only the initial stages of Si sublimation and C nucleation are discussed, which suggest that refined models could be addressed to support experimental optimization of epitaxial graphene growth on SiC substrates.

Acknowledgments

The author thanks his collaborators, H. Hibino, M. Nagase, H. Yamaguchi, S. Tanabe, and Y. Sekine. Almost all of the presented works were performed at NTT Basic Research Laboratories, Nippon Telegraph and Telephone Corporation facilities. This study is partially supported by KAKENHI projects 19310085, 21246006, 22310062, 22310086, 26289107 from JSPS. The first-principles calculations were performed with program package TAPP, which we have developed [99–102]. The atomic models shown in the view graphs are created by VESTA [103].

References

1. P. R. Wallace, *Phys. Rev.*, **71**, 622 (1947).
2. K. S. Novoselov, A. K. Geim, S. V. Morozov, D. Jiang, Y. Zhang, S. V. Dubonos, I. V. Grigorieva, and A. A. Firsov, *Science*, **306**, 666 (2004).
3. A. K. Geim and K. S. Novoselov, *Nat. Mater.*, **6**, 183 (2007).
4. A. K. Geim, *Science*, **324**, 1530 (2009).
5. A. H. Castro Neto, F. Guinea, N. M. R. Peres, K. S. Novoselov, and A. K. Geim, *Rev. Mod. Phys.*, **81**, 109 (2009).
6. A. K. Geim and I. V. Grigorieva, *Nature*, **499**, 419 (2013).
7. K. I. Bolotin, K. J. Sikes, Z. Jiang, M. Klima, G. Fudenberg, J. Hone, P. Kim, and H. L. Stormer, *Solid State Commun.*, **146**, 351 (2008).
8. M. Nagase, H. Hibino, H. Kageshima, and H. Yamaguchi, *Nanotechnology*, **19**, 495701 (2008).
9. M. Nagase, H. Hibino, H. Kageshima, and H. Yamaguchi, *Nanotechnology*, **20**, 445704 (2009).
10. J. Jobst, D. Waldmann, F. Speck, R. Hirner, D. K. Maude, Th. Seyller, and H. B. Weber, *Phys. Rev. B*, **81**, 195434 (2010).
11. H. Kageshima, H. Hibino, M. Nagase, Y. Sekine, and H. Yamaguchi, *Appl. Phys. Express*, **3**, 115103 (2010).
12. Y.-M. Lin, C. Dimitrakopoulos, K. A. Jenkins, D. B. Farmer, H.-Y. Chiu, A. Grill, and Ph. Avouris, *Science*, **327**, 662 (2010).
13. M. Nagase, H. Hibino, H. Kageshima, and H. Yamaguchi, *Appl. Phys. Express*, **3**, 045101 (2010).
14. Z. Zhang, C. Chen, X.-C. Zeng, and W. Guo, *Phys. Rev. B*, **81**, 155428 (2010).
15. H. Kageshima, H. Hibino, M. Nagase, Y. Sekine, and H. Yamaguchi, *Jpn. J. Appl. Phys.*, **50**, 070115 (2011).
16. S. Tanabe, Y. Sekine, H. Kageshima, M. Nagase, and H. Hibino, *Jpn. J. Appl. Phys.*, **50**, 04DN04 (2011).
17. S. Tanabe, M. Takamura, Y. Harada, H. Kageshima, and H. Hibino, *Appl. Phys. Express*, **5**, 125101 (2012).
18. K. Takase, S. Tanabe, S. Sasaki, H. Hibino, and K. Muraki, *Phys. Rev. B*, **86**, 165435 (2012).
19. M. Nagase, H. Hibino, H. Kageshima, and H. Yamaguchi, *Appl. Phys. Express*, **6**, 055101 (2013).

20. S. Tanabe, M. Takamura, Y. Harada, H. Kageshima, and H. Hibino, *Jpn. J. Appl. Phys.*, **53**, 04EN01 (2014).
21. I. Forbeaux, J.-M. Themlin, and J.-M. Debever, *Phys. Rev. B*, **58**, 16396 (1998).
22. A. Fissel, *Phys. Rep.*, **379**, 149 (2003).
23. C. Berger, Z. Song, T. Li, X. Li, A. Y. Ogbazghi, R. Feng, Z. Dai, A. N. Marchenkov, E. H. Conrad, P. N. First, and W. A. de Heer, *J. Phys. Chem. B*, **108**, 19912 (2004).
24. J. Hass, R. Feng, T. Li, X. Li, Z. Zong, W. A. de Heer, P. N. First, E. H. Conrad, C. A. Jeffrey, and C. Berger, *Appl. Phys. Lett.*, **89**, 143106 (2006).
25. T. Ohta, A. Bostwick, Th. Seyller, and E. Rotenberg, *Science*, **313**, 951 (2006).
26. H. Hibino, H. Kageshima, and M. Nagase, *J. Phys. D*, **43**, 374005 (2010).
27. H. Hibino, S. Tanabe, and H. Kageshima, *J. Phys. D*, **45**, 154008 (2012).
28. Y. Tateyama, T. Ogitsu, K. Kusakabe, and S. Tsuneyuki, *Phys. Rev. B*, **54**, 14994 (1996).
29. K. Oura, V. G. Lifshits, A. A. Saranin, A. V. Zotov, M. Katayama, *Surface Science: An Introduction* (Springer, Berlin), ISBN 3-540-00545-5, (2003).
30. Y. Homma, H. Hibino, T. Ogino, and N. Aizawa, *Phys. Rev. B*, **55**, R10237 (1997).
31. H. Kageshima, H. Hibino, and M. Nagase, *Mater. Sci. Forum*, **645–648**, 597 (2009).
32. H. Kageshima, H. Hibino, M. Nagase, and H. Yamaguchi, *Appl. Phys. Express*, **2**, 065502 (2009).
33. H. Kageshima, H. Hibino, H. Yamaguchi, and M. Nagase, *Jpn. J. Appl. Phys.*, **50**, 095601 (2011).
34. H. Kageshima, H. Hibino, and S. Tanabe, *J. Phys. Condens. Matter*, **24**, 314215 (2012).
35. H. Kageshima, H. Hibino, H. Yamaguchi, and M. Nagase, *Phys. Rev. B*, **88**, 235405 (2013).
36. H. Kageshima, H. Hibino, H. Yamaguchi, and M. Nagase, *Mater. Sci. Forum*, **778–780**, 1150 (2014).
37. H. Kageshima and H. Hibino, *Proceedings of 10th International Symposium on Atomic Level Characterizations for New Materials and Devices* '15, 28p-B-7 (2015).
38. M. Inoue, H. Kageshima, Y. Kangawa, and K. Kakimoto, *Phys. Rev. B*, **86**, 085417 (2012).

39. J. Hass, F. Varchon, J. E. Millan-Otoya, M. Sprinkle, N. Sharma, W. A. de Heer, C. Berger, P. N. First, L. Magaud, and E. H. Conrad, *Phys. Rev. Lett.*, **100**, 125504 (2008).
40. F. Varchon, P. Mallet, L. Magaud, and J.-Y. Veuillen, *Phys. Rev. B*, **77**, 165415 (2008).
41. U. Starke and C. Riedl, *J. Phys. Condens. Matter*, **21**, 134016 (2009).
42. C. Riedl, U. Starke, J. Bernhardt, M. Franke, and K. Heinz, *Phys. Rev. B*, **76**, 245406 (2007).
43. G. M. Rutter, N. P. Guisinger, J. N. Crain, E. A. A. Jarvis, M. D. Stiles, T. Li, P. N. First, and J. A. Stroscio, *Phys. Rev. B*, **76**, 235416 (2007).
44. S. Y. Zhou, G.-H. Gweon, A. V. Fedorov, P. N. First, W. A. de Heer, F. Guinea, A. H. Castro Neto, and A. Lanzara, *Nat. Mater.*, **6**, 770 (2007).
45. H. Hibino, H. Kageshima, F. Maeda, M. Nagase, Y. Kobayashi, Y. Kobayashi, and H. Yamaguchi, *e-J. Surf. Sci. Nanotechnol.*, **6**, 107 (2008).
46. H. Hibino, H. Kageshima, F. Maeda, M. Nagase, Y. Kobayashi, and H. Yamaguchi, *Phys. Rev. B*, **77**, 075413 (2008).
47. H. Hibino, H. Kageshima, F.-Z. Guo, F. Maeda, M. Kotsugi, and Y. Watanabe, *Appl. Surf. Sci.*, **254**, 7596 (2008).
48. P. Lauffer, K. V. Emtsev, R. Graupner, Th. Seyller, and L. Ley, *Phys. Rev. B*, **77**, 155426 (2008).
49. E. Rotenberg, A. Bostwick, T. Ohta, J. L. McChesney, Th. Seyller, and K. Horn, *Nat. Mater.*, **7**, 258 (2008).
50. S. Y. Zhou, D. A. Siegel, A. V. Fedorov, F. El Gabaly, A. K. Schmid, A. H. Castro Neto, D.-H. Lee, and A. Lanzara, *Nat. Mater.*, **7**, 259 (2008).
51. H. Hibino, H. Kageshima, M. Kotsugi, F. Maeda, F.-Z. Guo, and Y. Watanabe, *Phys. Rev. B*, **79**, 125437 (2009).
52. H. Hibino, S. Mizuno, H. Kageshima, M. Nagase, and H. Yamaguchi, *Phys. Rev. B*, **80**, 085406 (2009).
53. K. Nakatsuji, Y. Shibata, R. Niikura, and F. Komori, *Phys. Rev. B*, **82**, 045428 (2010).
54. Y. Qi, S. H. Rhim, G. F. Sun, M. Weinert, and L. Li, *Phys. Rev. Lett.*, **105**, 085502 (2010).
55. U. Starke, *Phys. Sol. Stat.*, **246**, 1569 (2009).
56. A. Caboni, N. Camara, E. Pausas, N. Mestres, and P. Godignon, *Mater. Sci. Forum*, **679–680**, 781 (2011).
57. N. Camara, B. Jouault, A. Caboni, A. Tiberj, P. Godignon, and J. Camassel, *Nanosci. Nanotechnol. Lett.*, **3**, 49 (2011).

58. B. Jabakhanji, N. Camara, A. Caboni, C. Consejo, B. Jouault, P. Godignon, and J. Camassel, *Mater. Sci. Forum*, **711**, 235 (2012).

59. J.-J. Lin, L.-W. Guo, Y.-P. Jia, L.-L. Chen, W. Lu, J. Huang, and X.-L. Chen, *Chin. Phys. B*, **22**, 016301 (2013).

60. M. Ostler, I. Deretzis, S. Mammadov, F. Giannazzo, G. Nicotra, C. Spinella, Th. Seyller, and A. La Magna, *Phys. Rev. B*, **88**, 085408 (2013).

61. K. V. Emtsev, F. Speck, Th. Seyller, L. Ley, and J. D. Riley, *Phys. Rev. B*, **77**, 155303 (2008).

62. C. Riedl, C. Coletti, T. Iwasaki, A. A. Zakharov, and U. Starke, *Phys. Rev. Lett.*, **103**, 246804 (2009).

63. S. Tanabe, Y. Sekine, H. Kageshima, and H. Hibino, *Jpn. J. Appl. Phys.*, **51**, 02BN02 (2012).

64. F. Hiebel, P. Mallet, F. Varchon, L. Magaud, and J.-Y. Veuillen, *Phys. Rev. B*, **78**, 153412 (2008).

65. F. Hiebel, P. Mallet, L. Magaud, and J.-Y. Veuillen, *Phys. Rev. B*, **80**, 235429 (2009).

66. L. Magaud, F. Hiebel, F. Varchon, P. Mallet, and J.-Y. Veuillen, *Phys. Rev. B*, **79**, 161405(R) (2009).

67. F. Hiebel, P. Mallet, J.-Y. Veuillen, and L. Magaud, *Phys. Rev. B*, **83**, 075438 (2011).

68. F. Hiebel, L. Maggaud, P. Mallet, and J.-Y. Veuillen, *J. Phys. D*, **45**, 153003 (2012).

69. T. Ohta, A. Bostwick, J. L. McChesney, Th. Seyller, K. Horn, and E. Rotenberg, *Phys. Rev. Lett.*, **98**, 206802 (2007).

70. S. Tanabe, Y. Sekine, H. Kageshima, M. Nagase, and H. Hibino, *Appl. Phys. Express*, **3**, 075102 (2010).

71. S. Tanabe, Y. Sekine, H. Kageshima, M. Nagase, and H. Hibino, *Phys. Rev. B*, **84**, 115458 (2011).

72. A. Bostwick, K. V. Emtsev, K. Horn, E. Huwald, L. Ley, J. L. McChesney, T. Ohta, J. D. Riley, E. Rotenberg, F. Speck, and Th. Seyller, *Adv. Solid State Phys.*, **47**, 159 (2008).

73. J. B. Hannon and T. M. Tromp, *Phys. Rev. B*, **77**, 241404(R) (2008).

74. M. L. Bolen, S. E. Harrison, L. B. Biedermann, and M. A. Capano, *Phys. Rev. B*, **80**, 115433 (2009).

75. V. Borovikov and A. Zangwill, *Phys. Rev. B*, **80**, 121506(R) (2009).

76. R. M. Tromp and J. B. Hannon, *Phys. Rev. Lett.*, **102**, 106104 (2009).

77. T. Ohta, N. C. Bartelt, S. Nie, K. Thürmer, and G. L. Kellogg, *Phys. Rev. B*, **81**, 121411(R) (2010).

78. S. Tanaka, K. Morita, and H. Hibino, *Phys. Rev. B*, **81**, 041406(R) (2010).

79. K. V. Emtsev, A. Bostwick, K. Horn, J. Jobst, G. L. Kellogg, L. Ley, J. L. McChesney, T. Ohta, S. A. Reshanov, J. Röhrl, E. Rotenberg, A. K. Schmid, D. Waldmann, H. B. Weber, and Th. Seyller, *Nat. Mater.*, **8**, 203 (2009).

80. M. Kusunoki, M. Rokkaku, and T. Suzuki, *Appl. Phys. Lett.*, **77**, 531 (2000).

81. Z. G. Cambaza, G. Yushinb, S. Osswalda, V. Mochalina, and Y. Gogotsi, *Carbon*, **46**, 841 (2008).

82. S. Kim, J. Ihm, H. J. Choi, and Y.-W. Son, *Phys. Rev. Lett.*, **100**, 176802 (2008).

83. A. Mattausch and O. Pankratov, *Phys. Rev. Lett.*, **99**, 076802 (2007).

84. F. Varchon, R. Feng, J. Hass, X. Li, B. Ngoc Nguyen, C. Naud, P. Mallet, J.-Y. Veuillen, C. Berger, E. H. Conrad, and L. Magaud, *Phys. Rev. Lett.*, **99**, 126805 (2007).

85. J. B. Hannon, M. Copel, and R. M. Tromp, *Phys. Rev. Lett.*, **107**, 166101 (2011).

86. Y. Kitou, E. Makino, K. Ikeda, M. Nagakubo, and S. Onda, *Mater. Sci. Forum*, **527–529**, 107 (2006).

87. Y. Kangawa, T. Ito, Y. S. Hiraoka, A. Taguchi, K. Shiraishi, and T. Ohachi, *Surf. Sci.*, **507–510** (2002) 285.

88. Yu. Goldberg, M. E. Levinshtein, and S. L. Rumyantsev, in *Properties of Advanced Semiconductor Materials GaN, AlN, SiC, BN, SiC, SiGe*, edited by M. E. Levinshtein, S. L. Rumyantsev, and M. S. Shur (John Wiley & Sons, Inc., New York, 2001), pp. 93–148.

89. D. P. Kosimov, A. A. Dzhurakhalov, and F. M. Peeters, *Phys. Rev. B*, **78**, 235433 (2008).

90. D. P. Kosimov, A. A. Dzhurakhalov, and F. M. Peeters, *Phys. Rev. B*, **81**, 195414 (2010).

91. M. Inoue, Y. Kangawa, H. Kageshima, and K. Kakimoto, *17th International Conference on Crystal Growth and Epitaxy* (*ICCGE-17*), Warsaw, Poland, ThO3-T07-2 (August 15, 2013).

92. M. Inoue, Y. Kangawa, K. Wakabayashi, H. Kageshima, and K. Kakimoto, *Jpn. J. Appl. Phys.*, **50**, 038003 (2011).

93. G. F. Sun, Y. Liu, S. H. Rhim, J. F. Jia, Q. K. Xue, M. Weinert, and L. Li, *Phys. Rev. B*, **84**, 195455 (2011).

94. H. Hibino, H. Kageshima, M. Nagase, and H. Yamaguchi, *Abstracts of 2010 Spring Meeting of Japanese Society of Applied Physics*, 18p-TE-5 (2010), in Japanese.

95. G. Lupina, J. Kitzmann, M. Lukosius, J. Dabrowski, A. Wolff, and W. Mehr, *Appl. Phys. Lett.*, **103**, 263101 (2013).

96. J. Ristein, S. Mammadov, and Th. Seyller, *Phys. Rev. Lett.*, **108**, 246104 (2012).

97. F. Ming and A. Zangwill, *Phys. Rev. B*, **84**, 115459 (2011).

98. F. Ming and A. Zangwill, *J. Phys. D*, **45**, 154007 (2012).

99. J. Yamauchi, M. Tsukada, S. Watanabe, and O. Sugino, *Phys. Rev. B*, **54**, 5586 (1996).

100. H. Kageshima and K. Shiraishi, *Phys. Rev. B*, **56**, 14985 (1997).

101. S. Yabuuchi, H. Kageshima, Y. Ono, M. Nagase, A. Fujiwara, and E. Ohta, *Phys. Rev. B*, **78**, 045307 (2008).

102. H. Kageshima and M. Kasu, *Jpn. J. Appl. Phys.*, **48**, 111602 (2009).

103. K. Momma and F. Izumi, *J. Appl. Crystallogr.*, **44**, 1272 (2011).

Chapter 6

Epitaxial Graphene on SiC from the Viewpoint of Planar Technology

Gemma Rius[a] and Philippe Godignon[b]
[a]*NEMS and Nanofabrication Group, Institut de Microelectrònica de Barcelona, IMB-CNM-CSIC, Campus UAB, 08193, Bellaterra, Catalunya, Spain*
[b]*Power Devices Group, Institut de Microelectrònica de Barcelona, IMB-CNM-CSIC, Campus UAB, 08193, Bellaterra, Catalunya, Spain*
gemma.rius@imb-cnm.csic.es

Early from the very demonstration of its superior, as well as unique, electronic properties, graphene has been investigated as one of the most promising materials ever, and surely nowadays. Particularly, graphene is considered, and has already demonstrated, to be a breakthrough addition to the electronic materials family, exhibiting superior performance potential for a number of electronic and sensing devices. Yet, the full exploitation of graphene as a crucial element of electronic devices faces some technical difficulties derived from its manipulation and the fabrication processes that have to be used. Limitations include high sensitivity to the environment, corrugation, wrinkles, impurities, etc. Some of those limitations are currently intended to be addressed by the combination of graphene with other 2D materials, such as hexagonal BN and MoS_2.

Epitaxial Graphene on Silicon Carbide: Modeling, Devices, and Applications
Edited by Gemma Rius and Philippe Godignon

ISBN 978-981-4774-20-8 (Hardcover), 978-1-315-18614-6 (eBook)
www.panstanford.com

Generally speaking, graphene, by definition a single atom–thick layer of densely packed C atoms, possesses a plethora of exceptional mechanical, optical, chemical, and electronic characteristics. Seen as a new material, it would enable infinite number of novel applications in many industrial sectors. Specifically for electronics, its pristine structure provides massless Dirac fermions or ballistic transport properties at room temperature. Technologically, graphene's intrinsic 2D morphology is most appealing; it points toward easing the compatibility of this novel 2D material with the conventional fabrication processes and sequences of modern electronics, using 3D semiconductors and so-called (silicon) planar technology. This advantage is opposed to other 0D or 1D nanostructured carbon materials such as electronically relevant carbon nanotubes and fullerenes, which are also cheap, chemically simple materials with potential, highly interesting characteristics for electronics.

Briefly, integration of epitaxial graphene on SiC with the standard processing techniques for the planar fabrication of electronic devices is often considered the best candidate from the point of view of graphene synthesis. Among the available mechanisms and techniques for the synthesis or growth of graphene, (epitaxial) graphene deposition by thermal sublimation of Si on SiC wafers outstands for its simplicity and convenience. The advantages are, on the one hand, that SiC crystal acts as the support as well as it provides the carbon atoms (precursor) for the formation of graphene; on the other hand, SiC has a convenient functional property being either a doped or semi-insulating material, offering different configurations for electronic device architectures. As stated in the first chapter, SiC technology is mature today, and its compatibility with graphene processing is achievable as we will review in the next paragraphs.

In this chapter, after a brief introduction to the basic processing sequences and processes of silicon electronics planar technology, we gather some of our technical investigations and developments for the synthesis and fabrication of electronic devices based on epitaxial graphene on SiC. We frame our achievements within our peers' relevant works and summarize this contribution about technology with the present challenges and future work, which would be required to (i) upgrade epitaxial graphene on SiC and (ii) bring it up to the market to provide novel or superior electronic devices.

6.1 Semiconductors Planar Technology

Planar technology stands for a series of well-established treatments, processes, and sequences that allow the fabrication of (e.g., silicon-based) electronic devices. In other words, planar processing is how devices are made out of semiconductors. Basically, planar manufacturing relies on the combination of additive and subtractive treatments of materials with lithography techniques; patterning based on a 2D projection of the device elements by step-by-step and layer-upon-layer sequential processing. In the 95% case of modern electronics, the basic structure is a field-effect transistor (FET), mainly controlled through an isolated gate, the MOSFET.

As a paradigmatic, simplified example, fabrication of a MOSFET transistor (Fig. 6.1) starts upon a doped Si substrate, where two highly doped areas are defined by photolithography plus ion implantation. Then, subsequent lithography levels involving metal deposition and etching steps are used to pattern oxide/insulating layer, define gate and body contacts, as well as patterning of drain and source vias and contacts. Simple definitions of each of those processes and their main features are as follows. For a comprehensive review on micro/nanofabrication, we could recommend Ref. [1].

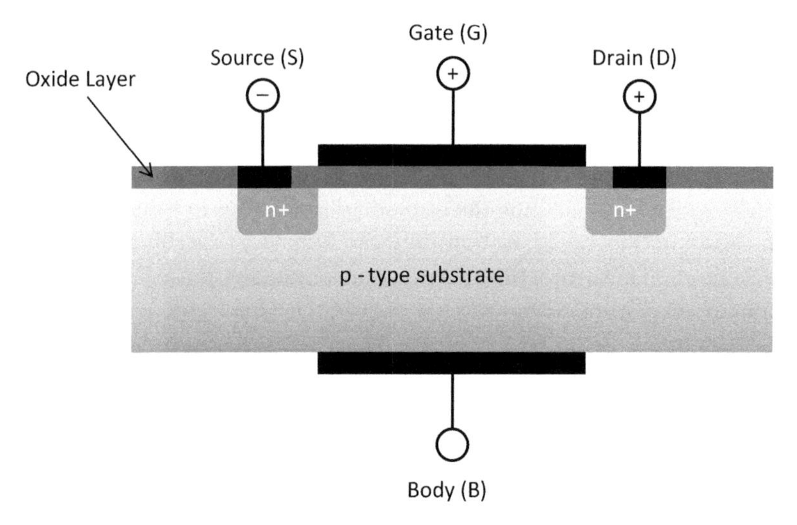

Figure 6.1 MOSFET transistor, the basic device of IC electronics, is manufactured by planar processing.

6.1.1 Semiconductors Surface and Bulk Treatments

A number of surface and bulk treatments are applied to the wafers of the semiconductor material as a part of the electronic device fabrication sequence, such as thermal oxidation and ion implantation. Thermal oxidation (of silicon) is an important step of the fabrication, as silicon oxide layers of precisely controlled thickness are used, e.g., as a mask for selective dopant introduction (ion implantation) or crucial element of the MOSFET gate (insulator). It is performed inside a furnace where O_2 or H_2O are introduced to react with Si and thus form the native SiO_2 oxide layer. O_2 and H_2O are used for thin and thick layer deposition in the so-called dry and wet oxidation, respectively. It is important to note that this natural high-quality oxide of Si is not seen in other semiconductors such as Ge, AsGa, SiC, GaN, diamond, which is one of the main forces of Si technology. Of course, graphene cannot be oxidized in the same way as Si since the graphene layer would be actually consumed during a wet or dry oxidation process.

Ion implantation consists in the introduction of impurity ions to alter the silicon bandgap (doping). It is done by generating ions of a given dopant impurity (B, As, P for Si) and accelerating them at determined high energies (up to MeV) in order to hit and controllably penetrate the silicon wafer. As a result of ion implantation, not only dopant impurities are introduced in interstitial position, but the Si atomic positions are displaced from their original position in the single crystal (damage). Consequently, high-temperature treatments are required to remove the damage, i.e., restore the crystallinity of Si, as well as to distribute the implanted impurities in substitutional position in the crystal, so that they can behave as carrier donors or acceptors. These practices are named annealing and diffusion or dopant activation.

Nonetheless, Si planar technology greatly owes its success to the (relative) processing ease and cost of (i) producing single-crystal silicon ingots and (ii) processing them as wafers of precisely cut thickness and diameter (iii) being able to polish them to attain a perfect mirror surface finish and (iv) the ease to grow a very high-quality natural dielectric SiO_2 for interface formation.

In contrast, the characteristic extreme hardness of SiC makes processes such as wafer cutting and mirror polishing remarkably expensive and difficult. Similarly, requirements for SiC single-crystal ingot growth are also commensurate, although great progress has been made recently with the production of 200 mm wafers. In general, the processing of SiC is more complex than Si, and the oxidation of SiC produces a low-quality interface SiO_2 oxide, with low oxidation rates and high density of defects. However, complex devices such as high-voltage power MOSFETs are currently produced and commercialized, thus competing with Si devices in markets such as electrical vehicle or renewable energies.

6.1.2 Patterning

The most important step of electronic device fabrication by planar technology is undoubtedly patterning. Although patterning techniques include mold processing and direct writing, lithography is commonly used. Patterning techniques can also be classified according to whether a resist media is used or not and their throughput (serial versus parallel techniques).

Photolithography or optical lithography is an easy, resist-based, parallel patterning technique, which still monopolizes electronic device fabrication. A mask is used to expose the resist selectively to light of a certain wavelength. After light irradiation, either the resist film is polymerized, and hence more difficult to dissolve (negative resist), or their polymer chains are broken down and thus easier to remove by a solvent (positive resist). Therefore, the resist can be selectively removed, according to the mask features, by a step called resist development.

Wavelength and its interaction with the resist material/film determine and limit the minimum feature size and density that can be defined. To reduce critical dimensions of the transistors (successive technologies generations of 48 nm, 35 nm, 28 nm, 14 nm...), high-energy light sources are being developed, greatly increasing processing cost and challenging conventional metal masks. Another limitation of the use of mask is the price of their fabrication, also increasing with shrinking feature dimensions, and limitations upon prototyping (no/poor patterning flexibility).

The next-generation lithography techniques, toward single-/double-digit nanometer-scale patterning capability, include electron lithography or nanoimprint. The most popular option for nanofabrication is definitely electron beam lithography (EBL), where an extremely fine energetic electron beam (~keV) is used to expose a thin resist layer. In terms of processing, it follows a chemical principle similar to photolithography. In addition to the possibility of patterning features with remarkable nanoscale precision, it is important that EBL provides absolute pattern flexibility and compatibility with photolithography-based fabrication processing (e.g., alignment or mix and match). However, as a serial technique, it introduces a limitation in terms of throughput.

6.1.3 Pattern Transfer

The features defined upon the layers of resist by patterning are not commonly useful by themselves, i.e., not as a part of the electronic devices, but as a mask for pattern transfer. Therefore, the transfer of the resist pattern to become part of the electronic device as a structural layer is needed. As mentioned in the introduction to the planar technology section, there are two ways to transfer the patterned features: (i) subtraction (etching) or (ii) addition (deposition) of thin-film materials.

Pattern transfer by etching consists of selective removal of matter based on the mask defined in the patterning step. Sometimes, etching can be applied directly using the resist as a mask; but, depending on etching selectivity and so on, to perform the etching by using other materials as a mask can be requisite. Etching techniques are commonly classified as dry or wet techniques, depending on the predominant principle they are based on. Dry techniques include reactive ion etching (RIE) or plasma etching and their main characteristic is that they can provide anisotropic transfer of the pattern, under optimized conditions. Instead, wet techniques are typically easier and cheaper as they simply consist in the immersion of the sample into chemical solutions. However, the transfer of the pattern is more or completely isotropic, which may compromise feature shape fidelity and size. A typical problem of wet etching is the under etching of layers below the photoresist mask. Process control is complex, especially with very sensitive layers, such as graphene.

Pattern transfer based on thin-film deposition includes a large variety of techniques as well as infinite possibilities in terms of materials. Structurally, crystalline, polycrystalline, and amorphous materials can be deposited, and their physical properties, such as electronic conduction, will also depend on their composition. In brief, metal, insulator, or semiconductor materials, and recently superconductors, can be incorporated. Common deposited materials include nitrides, oxides, and metals such as aluminum.

Apart from thermal oxidation of silicon, there are several other techniques used for material additive processing on top of the semiconductor; i.e., the thin-film deposition of materials, such as for silicon nitrides or metals. Sputtering and thermal evaporation are often referred to as physical deposition techniques and they consist in the extraction of atomic materials from a pure/compound target by the application of, e.g., an electric field or heating inside a vacuum chamber. These techniques are preferred for metal deposition and basically differ on their degree of conformal coverage. Particularly, thermal evaporation is preferred for nanofabrication by EBL as it provides better pattern transfer precision and eases the lift-off of the resist.

Chemical vapor deposition (CVD) consists in the growth of thin films based on gas-phase chemical reactions and precursors. CVD is routinely used to deposit films of polycrystalline silicon (poly-Si), SiO_2, and Si_3N_4. Variants of CVD include low-pressure chemical vapor deposition (LPCVD) and plasma-enhanced chemical vapor deposition (PECVD) processes. An interesting aspect of CVD is that dopant species can be introduced during the CVD deposition of Si, so that in situ doping of the deposited thin film is made available. The added advantage of PECVD is that it can be performed at lower temperature, which makes it compatible with the presence of temperature-sensitive layers such as metals and graphene. Atomic layer deposition is a more recent technique used for very thin-film deposition at relatively low temperature (typically $<300°C$), and with a great thickness precision.

In the following section, we will extend the discussion of the epitaxial growth of graphene (epitaxial graphene, EG) on SiC, understood also as an interesting combination of treatments/processes. Particularly, one can suggest that EG on SiC relates integrally to planar technology. EG on SiC can be seen as (i) a process, as it has been suggested in previous chapters, yet from a different vision; as well as (ii) epitaxy, but not strictly nor always; and even as (iii) an extreme case of thin-film deposition, i.e., of a single atom–thick film.

Connected to this, and only as a last remark, the continuous shrinking of feature dimensions as well as the introduction of new materials into conventional planar technology has been progressively demanding specific and advanced characterization techniques to assess the technical developments. Scanning electron microscopy (SEM), atomic force microscopy (AFM), high-resolution transmission electron microscopy (HR-TEM), as well as a number of spectroscopies are now consistently employed.

6.2 Making of EG on SiC Materials

Processing of graphene on SiC, e.g., to fabricate electronic devices, can take various forms and requires applying different approaches, either based on top-down or bottom-up strategies. Actually, the graphene layer often will not be used as a (as-grown) continuous/uniform layer. In most applications, one needs to define either rectangular stripes, squares, or any geometrical forms needed to build specific, discrete devices. This requires either to pattern the graphene (e.g., using a top-down strategy) or to directly grow the graphene with the required geometry (bottom-up approach) in order to form operative devices, such as the basic transistor or sensors.

From fabrication viewpoint, for applications where graphene is the only active or functional layer, SiC is often used merely as a mechanical substrate (support). In this case, it is preferable to use a semi-insulating SiC substrate to reduce the possible impact of the substrate on the electrical behavior and performance of the graphene-based system. Using SiC as a mere support, we are not constraint by the intrinsic issues of SiC technology and processing. Current SiC technology typically employs 4H-SiC off-axis-cut wafers. For EG growth, instead, one could in principle use any crystal orientation of the SiC to produce the graphene layer. Differently, if SiC layers are to become working elements of the graphene-based device, some technological considerations must be taken into account. Specifically, choices such as SiC face or the surface axis can be of relevance to get optimal electrical performances, as well as to limit the number of defects formed during SiC processing. Eventually, we may also want to isolate graphene from the SiC substrate. This is a common practice for CVD graphene and implies the transfer of the graphene layer onto another substrate; this required processing is often detrimental to graphene properties as well as costly.

6.2.1 SiC Substrate Preparation: Surface Polishing

As mentioned, the success of Si microelectronics owes great deal to the ease of the Si material, e.g., in terms of its wafer preprocessing for device fabrication. Remarkably, the established Si wafer technology allows obtaining mirror-like wafers very cost efficiently. Instead, as SiC is a significantly harder material than Si, polishing of SiC surface is technologically more complex. Indeed, a few companies are still controlling this process and significant differences can be found on the surface polishing quality depending on the wafer provider. This is an important aspect as technology strongly depends on the surface planarity, roughness, etc., especially when Schottky diodes [2] or MOSFET transistors are targeted [3]. For graphene growth, this aspect is also important when uniform, controlled deposition is attempted [4, 5].

SiC wafer preparation is typically based on four successive steps [3]; first, a grinding/lapping to planarize the wafer and get flat and parallel surfaces; then a mechanical polishing to decrease surface roughness. The third step is the more critical and consists of a chemical–mechanical polishing (CMP). In this process, the surface treatment is done by chemical reaction combined with a mechanical removal of the modified surface. The objective is to produce a surface with very low roughness, without scratches and avoiding subsurface damage. The parameters of this process strongly depend on the SiC wafer polytype, as well as doping and orientation. The last step consists of a chemical cleaning process to remove contaminants from the surface. Among the few companies offering very high-quality SiC polishing, we can mention Novasic, in France, and CEC, in Japan.

6.2.2 Standard Lithography

As deeply explained in previous chapters, silicon sublimation from the SiC causes a carbon-rich surface that nucleates forming an epitaxial graphene layer. The graphene growth rate was found to depend, for instance, on the specific polar face of the SiC crystal: Graphene can form continuous layers on the Si-face (Fig. 6.2a,b), while under certain growth conditions, graphene flakes can be obtained on the C-face of the surface [6]. As seen, in both cases, the resulting surface is composed of step bunched SiC with micrometer-wide terraces (Fig. 6.2a). All terraces run parallel to the (11–20) direction. Yet, SiC reconstruction resulting from graphene deposition processing is not necessarily elongated terraces with smooth edges regularly distributed, but its surface recrystallization depends on a complex set of variables and parameters. As an indication, for our standard technology for Si-face graphene deposition chip, size-elongated terraces have terrace widths ranging from sub-micron to a few tens of micron values, and step heights from a few to a hundred nanometers (Fig. 6.2a).

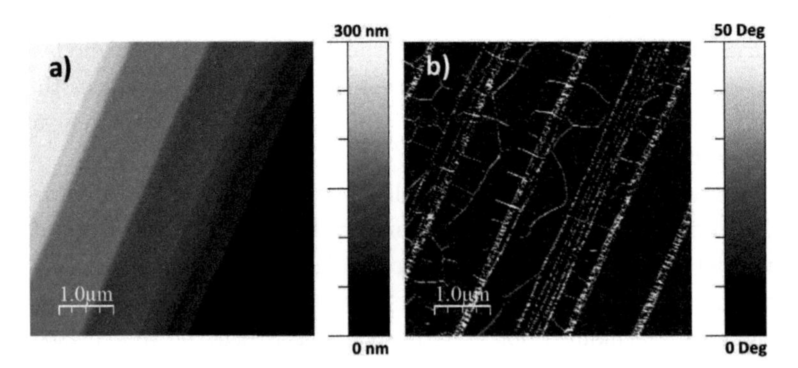

Figure 6.2 Exemplary AFM images of EG grown on the Si-face of 4H-SiC: (a) topography signal; (b) phase signal. Samples by courtesy of GraphNanotech.

Actually, it has been shown that the graphene grown on the terraces sidewall of the Si-face SiC provides poor quality for electronic devices. Therefore, to reach the optimal performance from the material, it has been suggested that lithography processes must be aligned with the terraces. In the case of as-grown isolated graphene flakes on the C-face, the flake must be localized and lithography must be done based on the exact flake position.

Provided the indeterminations or arbitrariness of the locations and shapes where and how graphene has to be patterned, for contacts and device shaping, conventional fabrication methods based on (multilevel) photolithography cannot be applied. Instead, EBL is the most convenient and reliable patterning method, as it is a direct writing technique with total flexibility in terms of pattern design.

Basically, the fabrication of, e.g., electronic devices relies on the combination of a few or more steps of the basic processing methods. For isolated graphene flakes, a simple electrical contacting is analogous to the electrical contacting of other nano-objects such as carbon nanotubes [7]. After the graphene deposition, alignment marks are patterned, for instance, by the sequence EBL on PMMA, deposition of a thin metal layer, and PMMA lift-off. Then, the exact coordinates of the isolated flakes are determined with respect to the alignment marks. Based on the coordinates and flake shape, a design is generated, such as two/four probes contacting. A second process of EBL, metal deposition, and resist removal to pattern the contact vias and pads. The principle and general hints on EBL and EBL-based fabrication, such as conventional resist materials preparation and typical exposure parameters, can be consulted elsewhere [8].

Fabrication processing issues related to epitaxial graphene on SiC can arise for a number of aspects. The stepped surface of SiC resulting from its high-temperature decomposition can be a problem for resist deposition, as thin PMMA 950k MW films (e.g., 100 nm) are commonly used. This topography question should also be considered for the metal deposition. An option is to use a thicker resist, e.g., a double layer such as PMMA 950k MW on PMMA 495k MW, which is also advantageous for metal contact deposition. Not only thicker metal films can be deposited, for a more robust device, but this kind of double layer eases the resist lift-off. In terms of EBL exposure, feature sizes and density are not typically the critical aspect for metal contacts, but some charging effects from semi-insulating SiC can be expected. Additionally, some difficulties on bonding and contact resistance are reported for operational electronic devices. High contact resistances and polymer residuals coming from the fabrication processing indeed appear. Typically, annealing treatments such as thermal annealing are enough to improve (reduce) contact resistance. The presence of PMMA residues, which are also found on CVD graphene merely due to transfer process, is more difficult to overcome, and it can be significantly detrimental to the electronic devices, particularly for fundamental studies where performance based on the characteristics of pristine graphene is needed. One solution based on high current cleaning has been proposed for PMMA on CVD graphene [9].

More complex graphene patterning implies additional steps and treatments. Specifically, for continuous films the use of masked etching to selectively remove graphene is required (Fig. 6.3). Similarly, the shaping of isolated flakes, as for instance the Hall bars in the following section, typically applies patterning too. An example and it fabrication conditions can be found in Ref. [10]. In addition to the possibility of more structural damage due to the extra manipulation, the main concern linked to the application of etching techniques is graphene edges doping and defects. This is particularly critical for spintronics applications. In cases where quantum confined graphene structures are sought after, reliable and precise-enough fabrication approaches are still needed. Nonetheless, most of the technical challenges are not exclusive for epitaxial graphene on SiC, but for any synthetic graphene such as CVD graphene.

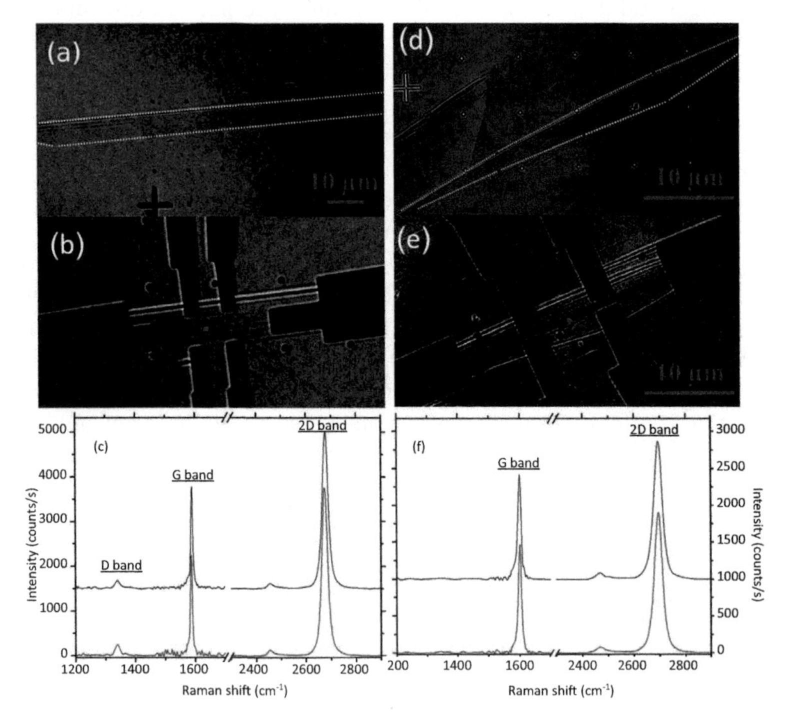

Figure 6.3 OM images (a–e) and Raman spectrum (c) and (d) for two quantum Hall bar structures of EG on the carbon face of 6H-SiC samples: (a, d) As-grown isolated single-layer graphene flakes; (b, e) patterned and metal-contacted graphene flakes; and (c, f) in Raman scattering.

We will next describe and demonstrate some of our (original) strategies based on planar processing. We have developed several instances of strategic synthesis as well as micro/nanofabrication of electronic devices applied to epitaxial graphene on SiC.

6.2.3 Typical Hall Bar Fabrication Process

We routinely prepared Hall bar structures using established EBL-based patterning. First, six-electrode Cr/Au (10/50 nm) contacts are deposited on the chosen isolated graphene ribbons. For doing so, EBL design is aligned respect to the as-deposited graphene ribbons and according to their exact coordinates, which have been determined on the basis of some additional metal marks (Fig. 6.3a,d). Then the ribbons can be selectively etched by oxygen plasma, in order to define the active graphene areas as a Hall bar structure (Fig. 6.3b,e). As the etching (shadow) mask, e.g., a thin PMMA layer patterned by EBL can be used.

Exemplary structures and their characterization are depicted in Fig. 6.3 for two ribbons obtained on two different substrates. In Fig. 6.3a,d, single-layer and bilayer graphene ribbons, respectively, were localized optically on the surface of the C-face of a semi-insulating SiC wafer. An optical view of the sample after contact deposition and etching of the Hall bar is presented in Fig. 6.3b,e. Raman spectra on two positions of each one of the patterned graphene Hall structures account for their crystal quality and uniform properties (Fig. 6.3c,f). Although the applied fabrication processes are well established, completely damage-less processing sequence is not guaranteed; therefore, alternative approaches for obtaining patterned graphene are proposed as follows.

6.2.4 Template Growth

The most conventional top-down graphene-based device technology is to form nanoribbons by resist patterning plus dry etching in an O_2 plasma. This step introduces defects and dangling bonds at the edges of graphene nanoribbons, decreasing their carrier(s) mobility [11]. Also, the polymethyl methacrylate (PMMA) and the chemical wet solutions used for patterning induce an unintentional doping in the graphene sheets, altering the device performances [9].

An alternative to the top-down standard technique, or to selectively deposit graphene, is to grow structured graphene by using wafer pre-patterning. In the case of CVD graphene, catalyst pre-patterning has been proposed and demonstrated to selectively grow graphene on microscale templates. Yet chemical doping, impurities, etc., remain as potential damage or limitations, provided the requisite chemical processing to obtain graphene on a dielectric substrate. On epitaxial graphene on SiC, template growth has also been reported. A template material able to withstand high temperature must be used as mask. Technical electronic materials include AlN (Fig. 6.4a) [12] or Si_3N_4.

Micron-size FLG can be produced (Fig. 6.4b) through a micro-patterned aluminum nitride (AlN) layer on top of an SiC sample (Fig. 6.4c). The mask consists of a few hundred thick sputtered AlN film. The deposited AlN prevents SiC surface decomposition, so also inhibits graphene formation except for the mask openings. In addition to an obvious relevance for industrial purpose, one of the main advantages of this technique is that the following steps for the fabrication of graphene electronic device, such as the dielectric layer used as top gate for the FET-like device, can be directly deposited, right after the graphene growth over the full wafer. In such a way, the patterned graphene ribbons are never subjected or in contact to chemical processing for the rest of the device processing. This strategy is advantageous or relevant only for non-sensing devices.

Giving more detail on this material properties for hard masking, AlN is currently used in SiC device technology as capping layer, protecting the implanted SiC surface during post-implantation annealing at high temperature [13, 14]. It is very stable at temperatures that silicon nitride (SiN) or silicon dioxide (SiO_2) cannot sustain. No reaction occurs between the AlN layer and the SiC substrate up to 1650°C [13]. It is also very easy to remove in a wet tetramethyl ammonium hydroxide etching. In Fig. 6.4, graphene material grown inside an AlN-patterned rectangle on the C-face 6H-SiC sample is shown [12]. A similar approach could also be applied successfully on the Si-face of 6H-SiC or 4H-SiC substrate samples. No doubt exists about the graphitic nature of the patterned area as FLG, since the Raman spectrum evidences that the signature peaks of sp^2 bonding are clearly registered (Fig. 6.4b). According to the I_D and I_G bands ratio, flakes dimensions are in the range of

30 nm in average, with a turbostratic arrangement. In comparison with unmasked growth on a similar substrate, graphene domain size is significantly smaller.

Additionally, results indicate that it might be better to align the AlN pattern along with the SiC crystal, for better graphene formation. In summary, this approach facilitates the graphene-based device process technology over the whole SiC wafer and opens the road to a new generation of graphene-based devices mixed with SiC-based devices on an integrated fashion.

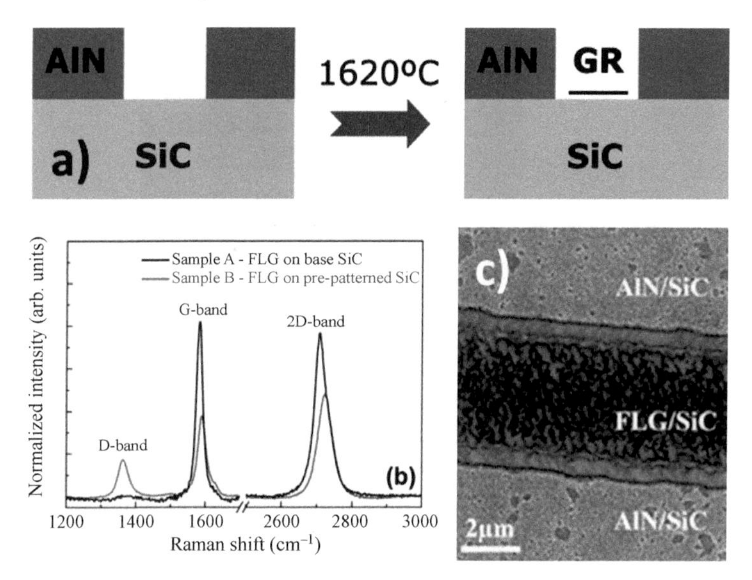

Figure 6.4 (a) Scheme of the AlN-based process for the synthesis of graphene ribbons (GR). (b) Raman spectra for FLG on bare SiC and AlN-mask pre-patterned FLG. (c) SEM image of a selectively synthesized FLG ribbon surrounded by AlN/SiC mask. Reprinted from Ref. [12], with permission from AIP Publishing.

Another approach proposed by Puybaret et al. in Ref. [15] is to use Si_3N_4 as template. Similar to the AlN experiment, the Si_3N_4 mask is deposited and patterned previous to the graphene growth. The main difference is that during the graphene growth, the Si_3N_4 layer will evaporate since on their deposition temperature, Si_3N_4 decomposition conditions are reached. Yet some specific conditions can be adjusted to grow graphene only on the patterned area, or to get monolayer graphene grown while blowing the Si_3N_4 areas as well as multilayer graphene in the patterned Si_3N_4 areas. This can be done playing with the stoichiometry of the Si_3N_4 and open different design options for future devices (Fig. 6.5).

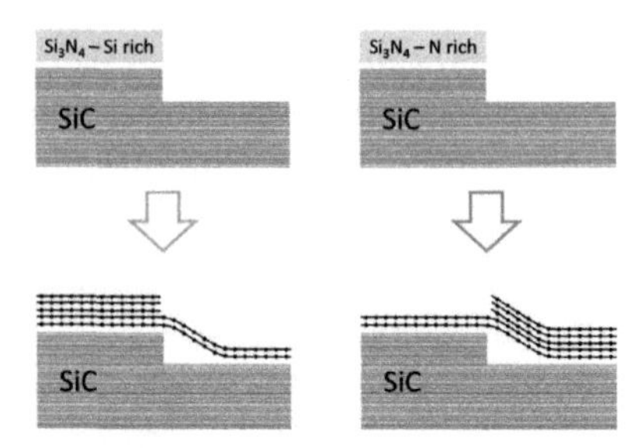

Figure 6.5 Diagrams for Si_3N_4-based synthesis of EG, as described in Ref. [15].

6.2.5 Nanoribbons

Interesting electronic properties arise from structuring graphene down to the single- and double-digit nanometer scale. Nanometer-width graphene ribbons exhibit bandgap opening arising from quantum confinement effects. The phenomenon has been observed in epitaxial graphene on SiC and combines with the carrier(s) high mobilities intrinsic to graphene on SiC. Theoretically calculated, bandgap energy separation is inversely proportional to the ribbon width and also depends on the edge type of the ribbons. Experimentally, nanoribbons obtained by a top-down approach, plasma etching after EBL, typically create rough edges. Non-ideal atomic-scale nanoribbon edges tend to act as electron scatters, and strong localization effects appear, which impedes the gap-opening formation. Accordingly, other ways to generate the nanoribbons are considered.

As an alternative to EBL, local anodic oxidation (LAO), a type of the scanning probe lithography (SPL) technique, has been proposed to structure and tune electronic properties of graphene [16], including epitaxial graphene on SiC ribbons [17]. Initially, and also traditionally, LAO has been extensively used on silicon and metal surfaces to pattern nanosized structures and built devices. The technique has also been evaluated on SiC [18], demonstrating that SiC can be oxidized in contact mode by applying a negative biasing of the probe with respect to the SiC sample. It allows forming an Si_xO_y dielectric layer with a high aspect ratio.

We have been able to modify electrically isolated graphene nanoribbons, specifically, by applying the same technique on single-layer EG samples grown on the C-face of 6H-SiC [17]. The results are illustrated in Fig. 6.6, where graphene ribbon has been first cut and then a rectangular section of the same graphene ribbon has been isolated from the surrounding layer by LAO. The first cut implies an increase in resistance due to the formation of a constriction; while the second cut fully isolates the area.

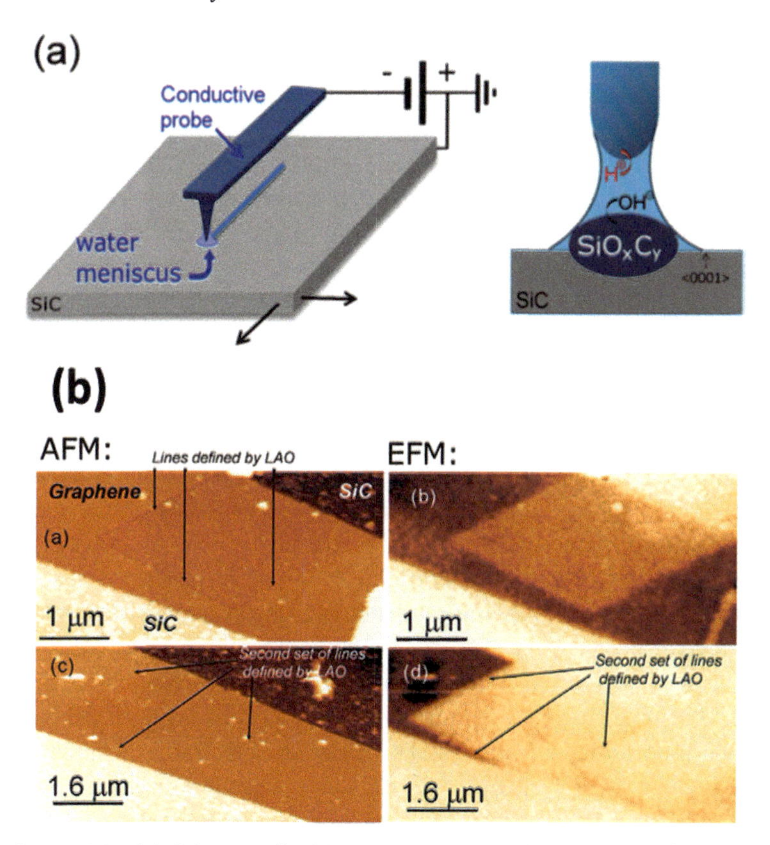

Figure 6.6 (a) Scheme of LAO–AFM nanopatterning on SiC substrates. Reprinted from Ref. [18], with the permission of AIP Publishing. (b) LAO–AFM nanopatterning of EG on SiC for modulation of graphene ribbon resistance. Reprinted with permission from Ref. [17], Copyright 2009, American Vacuum Society.

LAO experiments are performed with platinum-coated Si tips. Force constant is 4 N/m and resonance frequency is about 65 kHz.

The electric field resulting from the potential difference imposed between the AFM probe and the graphene flake together with the induced water meniscus provide environment condition locally; where activated oxygen atoms present in the air ambient can cut (etch) the graphene at the same time that slightly oxidize the SiC surface. Different from previously reported results on bare SiC [18], both negative and positive bias can be used successfully to etch graphene. Threshold voltages of +5.3 V or –6.5 V are necessary to obtain some surface modification. Qualitatively, the results indicate that several phenomena take place in addition to a pure anodic oxidation, such as also involving probably hydrogenation of the surfaces. In any case, it clearly demonstrates that efficient patterning of epitaxial graphene on SiC can be obtained by SPL and it can be applied to the patterning of functional structures or fabrication of operational devices with high resolution and low side bumps. As an example, with some more LAO technique optimization, we could propose creating a side gate using oxidized SiC as gate dielectric.

Another original method for graphene (nano)ribbons formation on SiC was proposed in Ref. [19] based on ion implantation. It relies on locally implanting Au or Si at (relatively) low energy, around 30 keV, and high dose (85×10^{16} at/cm^2) before the high-temperature graphitization process. Actually, in the implanted regions, graphene deposition appears at lower graphitization temperature than in SiC virgin areas. As a result, selective/patterned graphene growth can be obtained. However, further optimization of the process is needed in order to significantly improve the resulting graphene quality and reduce disorder.

Practically, so far the best way to generate high electronic performance graphene nanoribbons on SiC is to achieve selective growth according to the crystal facets of SiC substrates. Besides the on-axis facets, SiC has other crystal facets with low crystal indexes such as the SiC "sidewalls" that connect the terraces on the on-axis wafer surface. These low-index facets grouped as (110n) and (112n). Typically, the graphene epitaxial growth on these facets takes place at lower temperatures than on Si (0001) or C (000-1) faces, which is caused by a lower surface free energy and a weaker bonding of atomic Si. This peculiarity has been used to selectively grow high mobility graphene nanoribbons [20, 21]. The creation of facets is promoted by applying standard lithography aligned perpendicular to the

<00-10> direction in combination with dry etching (either SF_6 RIE or H_2 etching). Then, special growth parameters can be established to get selective growth, graphene formation only on the facet. In consequence, the nanoribbon width is determined by the etched sidewall depth/height. Once the graphene ribbon is grown, gate dielectric and metal contacts can be added to form FET transistors, as shown in the renowned Fig. 6.7 [20]. In the first experiments, mobility values of 2700 $cm^2/V\,s$ had been obtained and an on–off ratio of 10 in the gated transistors. More recently, very impressive results have been obtained by optimizing this technique. Fabrication of 40 nm wide nanoribbons-based transistors on the Si (0001) face, aligned along the <0-100> direction, are reported in Ref. [20]. Their electronic devices exhibit room-temperature ballistic conduction above 10 μm long paths, with carrier mobilities higher than 100,000 $cm^2/V\,s$ [22].

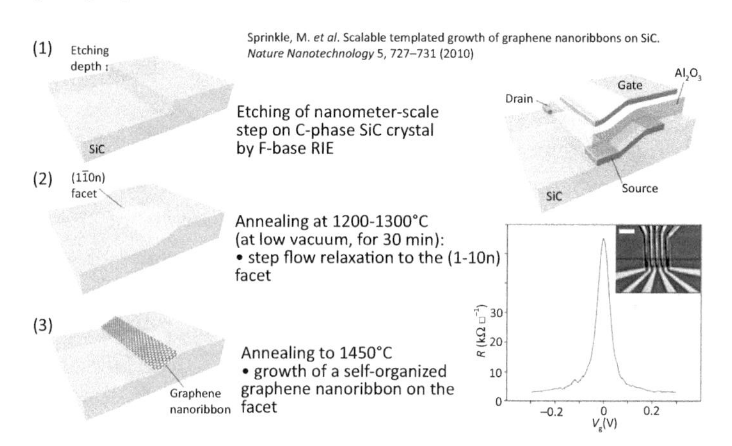

Figure 6.7 Scheme of the process steps for nanoribbon growth on SiC. Reprinted from Ref. [20], with permission from Macmillan Publishers Ltd.

6.3 Making of EG on Pre-processed SiC

6.3.1 Fabrication of Bottom-Gated Samples

SiC is a semiconductor material relevant for applications in the power devices industry. SiC device processing technologies have been developed in the last 20 years to produce power rectifiers and transistors. Today most of the technological steps are controlled enough to allow industrial production of high-reliability devices, with architectures as complex as those of power MOSFET transistors, which include more than 100 process steps. Technological steps are specifically p-type and n-type doping by implantation, trench etching, ohmic, and Schottky contact formation, MOS gate formation, etc.

Actually it is possible to use some of these technology developments to include as a processing of the SiC substrate before graphene growth. The main limitation is the high thermal budget of graphene deposition process, which may affect some process steps. For example, it will not be possible to use metals or other materials unless they can sustain the high temperature used during EG growth. Fabrication processing after graphene growth is also possible, but compatibility study is needed, mainly, to avoid using intensive extra process steps, which may damage the graphene structurally and chemically.

In this strategy, one of the possible approaches is to create a back gate by dopant incorporation into the SiC, i.e., to form a buried highly doped layer, which will act as the electrode of the gate. Buried gates have been demonstrated in Ref. [23] using high-energy nitrogen implantation. The concept of the buried gate device and the implantation profile inside the SiC proposed in Ref. [23] is shown in Fig. 6.8. The low nitrogen concentration at the SiC surface ensures the realization of an insulating barrier between the graphene and a conducting layer, which will act as the gate. The conducting layer is located 100–200 nm below the surface, as indicated by the maximum of the N concentration. The buried layer in the TEM cross section can be clearly seen in Fig. 6.8. The N implantation is performed through an oxide mask, which is removed before EG growth. This approach allows implementing all types of transistor or structure design. This concept can be used both for Si-face or C-face, as well as on off-axis and on-axis substrates.

In the example shown in Fig. 6.8, we used as starting material the C-face of a semi-insulating on-axis 6H-SiC substrate from the company CREE, with residual concentration 10^{15} cm^{-3}. For the ion implantation, we used a low-energy EATON 4200 NV implanter, with the implantation performed at the full wafer scale. The implanted specie was nitrogen, in other words, the standard donor used for SiC. The acceleration energy was 200/keV and a dose of 5×10^{13} cm^{-2} was used.

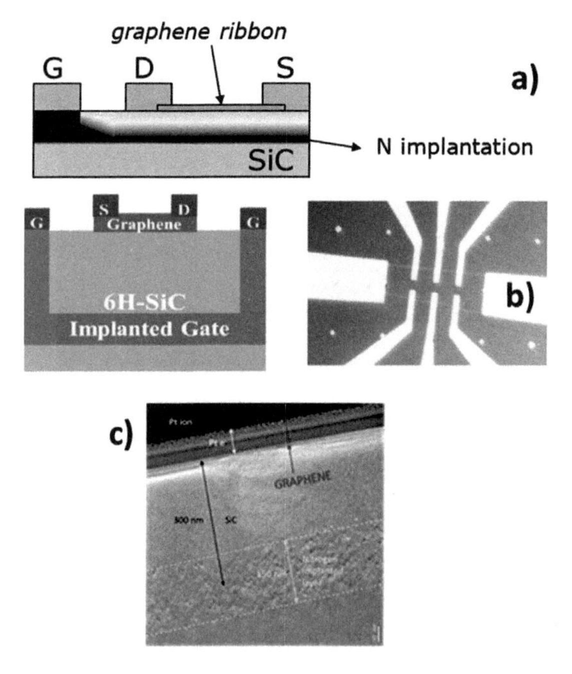

Figure 6.8 Bottom-gate EG-SiC devices based on ion implantation for Hall effect studies. Reprinted from Ref. [23], with the permission of AIP Publishing.

For device operation, the fabrication includes the processing to enable biasing the implanted conductive layer. This can be obtained by an additional implantation at a higher dose and lower energy performed in the corners of the transistor. After implantation, the substrate must be annealed in our case at 1700°C during 40 min in a confined atmosphere, which conditions are used to simultaneously graphitize. Monolayer graphene ribbons with typical widths of 5–10 μm and lengths up to 300 μm have been obtained by this method on the C-face of SiC and contacted using EBL patterning plus Cr/Au deposition of invasive contacts. We present exemplary results obtained in Fig. 6.8. A similar process can be used on the Si-face of an SiC substrate, if an additional step to pattern the graphene ribbons of the transistors is used.

This kind of technology is currently commercially available. Graphene Nanotech is offering this graphene with specification as follows: measured mobilities of 12.000 cm^2/Vs at 3 K and more than 3000 cm^2/Vs at room temperature. Quantum Hall effect is also observed in these samples at a reasonable magnetic field (Fig. 6.9).

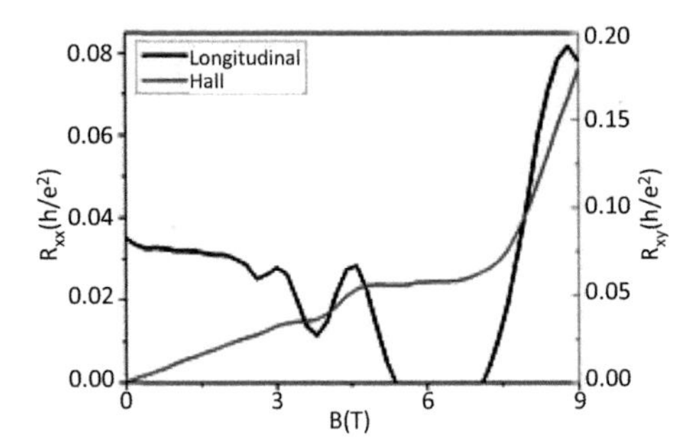

Figure 6.9 Observation of quantized resistance in EG–SiC devices based on ion implantation bottom-gating. Reprinted from Ref. [23], with the permission of AIP Publishing.

Similar process and performance results have been presented in parallel on the Si-face 6H-SiC by Friedrich-Alexander University, Erlangen [24]. In addition, the same group developed a transistor technology combining graphene with silicon carbide (SiC) [25]. In this work, a MESFET transistor is developed and processed on an SiC substrate with an epilayer. The MESFET channel is created inside the SiC, and the graphene acts as metal for the ohmic contacts of drain and source, and as gate metal. In this case, the MESFET fabrication includes isolation etching, source/drain nitrogen implantation. The graphene is used for the interconnections, and more complex logic circuits could be build using this concept of sequenced processing (planar technology).

Technologically, one of the key aspects to expand the potential applications of graphene grown on SiC would be the possibility to transfer the graphene layers on other substrates, as it is routinely done with CVD graphene grown on metals. As SiC cannot be chemically etched, such as copper or nickel, purely mechanical or dry transfer must be considered. In addition, the strong binding of graphene on SiC surface, especially on Si-face, makes this mechanical transfer highly complex if based on mechanical exfoliation, e.g., assisted by thermal tape, stressor layers, etc. However, recent experiments shown that under certain conditions, high-quality SLG graphene layers grown on SiC can be transferred. The work done by Kim et al. [26] is particularly interesting as they used the strain generated by an adhesive stressor layer, concretely nickel. With an adhesive tape, it is possible the remove the Ni layer with the graphene from the SiC surface thanks to the difference of binding energies between the Ni layer/graphene and the energy of the graphene/SiC interface. A previous work was proposed [27] using thermal release tape on multilayer graphene grown on the C-face of the 6H-SiC substrate. In this case, they take profit of the lower binding energy of graphene/graphene interface compared to SiC/graphene. Good-quality layers were also successfully transferred on Si substrates.

Acknowledgments

G. R. acknowledges partial financial support from Catalan Government Beatriu de Pinós Fellowship 2014BP-B-00207.

References

1. B. Ziaie, A. Baldi, and M. Atashbar, Introduction to micro/nanofabrication, in *Springer Handbook of Nanotechnology*, B. Bhushan (Ed.), pp. 231–269, Springer-Verlag Berlin Heidelberg (2007).
2. P. Godignon, R. Perez, D. Tournier, N. Mestres, H. Mank, and D. Turover, SiC power diodes improvement by fine surface polishing, *MRS Proceedings*, **815**, J5–12 (2004).
3. W. J. Choyke, H. Matsunami, and G. Pensl (Eds.). *Silicon Carbide: Recent Major Advances*. Springer Science & Business Media (2013).
4. G. Rius, N. Mestres, Y. Tanaka, H. Miyazaki, O. Eryu, and P. Godignon, *Mater. Sci. Forum*, **778**, 1158–1161 (2014).
5. G. Rius, N. Mestres, O. Eryu, and P. Godignon, *Mater. Sci. Forum*, **821**, 953–956 (2015).
6. N. Camara, G. Rius, J.-R. Huntzinger, A. Tiberj, L. Magaud, N. Mestres, P. Godignon, and J. Camassel, *Appl. Phys. Lett.*, **93**(26), 263102 (2008).

7. G. Rius, A. Verdaguer, F. A. Chaves, I. Martin, P. Godignon, E. Lora-Tamayo, D. Jimenez, and F. Pérez-Murano, *Microelectron. Eng.*, **85**(5), 1413–1416 (2008).
8. G. Rius, *Electron Beam Lithography for Nanofabrication*. Universitat Autònoma de Barcelona (2008).
9. J. Moser, A. Barreiro, and A. Bachtold, *Appl. Phys. Lett.*, **91**, 163513 (2007).
10. F. Molitor, J. Güttinger, C. Stampfer, D. Graf, T. Ihn, and K. Ensslin, *Phys. Rev. B*, **76**, 245426 (2007).
11. K. I. Bolotin, K. J. Sikes, Z. Jiang, M. Klima, G. Fudenberg, J. Hone, P. Kim, and H. L. Stormer, *Solid State Commun.*, **146**, 351 (2008).
12. N. Camara, G. Rius, J.-R. Huntzinger, A. Tiberj, N. Mestres, P. Godignon, and J. Camassel, *Appl. Phys. Lett.*, **93**, 123503 (2008).
13. M. A. Derenge, K. A. Jones, K. W. Kirchner, M. H. Ervin, T. S. Zheleva, S. Hullavarad, and R. D. Vispute, *Solid State Electron.*, **48**, 1867 (2004).
14. K. A. Jones, M. A. Derenge, M. H. Ervin, P. B. Shah, J. A. Freitas, R. D. Vispute, R. P. Sharma, and G. J. Gerardi, *Phys. Status Solidi A*, **201**, 486 (2004).
15. R. Puybaret, J. Hankinson, J. Palmer, C. Bouvier, A. Ougazzaden, P. L. Voss, C. Berger, and W. A. de Heer, *J. Phys. D Appl. Phys.*, **48**(15), 152001 (2015).
16. R. Garcia, A. W. Knoll, and E. Riedo, *Nat. Nanotechnol.*, **9**(8), 577–587 (2014).
17. G. Rius, N. Camara, P. Godignon, and F. Perez-Murano, *J. Vac. Sci. Technol. B*, **27**(6), 31493152 (2009).
18. M. Lorenzoni and B. Torre, *Appl. Phys. Lett.*, **103**, 163109 (2013).
19. S. Tongay, M. Lemaitre, J. Fridmann, A. F. Hebard, B. P. Gila, and B. R. Appleton, *Appl. Phys. Lett.*, **100**, 073501 (2012).
20. M. Sprinkle, M. Ruan, Y. Hu, J. Hankinson, M. Rubio-Roy, B. Zhang, X. Wu, C. Berger, and W. A. de Heer, *Nat. Nanotechnol.*, **5**, 727–731 (2010).
21. Q. Huang, J. Kim, G. Ali, and S. Oh Cho, *Adv. Mater.*, **25**, 1144–1148 (2013).
22. J. Baringhaus, M. Ruan, F. Edler, A.Tejeda, M. Sicot, A. Taleb-Ibrahimi, A.-P. Li, Z. Jiang, E. H. Conrad, C. Berger, C. Tegenkamp, and W. A. de Heer, *Nature*, **506**, 349354 (2014).
23. B. Jouault, N. Camara, B. Jabakhanji, A. Caboni, C. Consejo, P. Godignon, D. K. Maude, and J. Camassel, *Appl. Phys. Lett.*, **100**, 052102 (2012).

24. D. Waldmann, J. Jobst, F. Speck, T. Seyller, M. Krieger, and H. Weber, *Nat. Mater.*, **10**(5), 357–360 (2011).

25. S. Hertel, D. Waldmann, J. Jobst, A. Albert, M. Albrecht, S. Reshanov, A. Schöner, M. Krieger, and H. B. Weber, *Nat. Commun.*, **3**, 957 (2012).

26. J. Kim, H. Park, J. Hannon, S. Bedell, K. Fogel, D. Sadana, and C. Dimitrakopulos, *Science*, **342**, 833–836 (2013).

27. J. D. Caldwell, T. J. Anderson, J. C. Culbertson, G. G. Jernigan, K. D. Hobart†, F. J. Kub, M. J. Tadjer, J. L. Tedesco, J. K. Hite, M. A. Mastro, R. L. Myers-Ward, C. R. Eddy Jr., P. M. Campbell, and D. Kurt Gaskill, *ACS Nano*, **4**(2), 1108–1114 (2010).

Chapter 7

Beauty of Quantum Transport in Graphene

Benoit Jouault,[a] Félicien Schopfer,[b] and Wilfrid Poirier[b]
[a]*Laboratoire Charles Coulomb, CNRS-Université de Montpellier, Place Eugène Bataillon, 34095 Montpellier, France*
[b]*Laboratoire National de Métrologie et d'Essais, avenue Roger Hennequin, 78197 Trappes, France*
benoit.jouault@umontpellier.fr

7.1 Introduction

The famous integer quantum Hall effect (QHE) [58], discovered in 1980 [34], is one of the most fascinating quantum effects existing in condensed matter physics. When a Hall bar, based on semiconductor heterostructures in which the electrons can move only in the two dimensions defining the plane of the Hall bar, is cooled down to low temperature (1.5 K) and placed in a strong perpendicular magnetic flux density (10 T), the Hall effect breaks down and the QHE manifests itself. The Hall resistance R_H develops steps and plateaus. The value of the Hall resistance is quantized on the plateaus and given by $R_H = R_K/N$, where N is an integer, R_K is the von Klitzing constant,

Epitaxial Graphene on Silicon Carbide: Modeling, Devices, and Applications
Edited by Gemma Rius and Philippe Godignon

ISBN 978-981-4774-20-8 (Hardcover), 978-1-315-18614-6 (eBook)
www.panstanford.com

theoretically equal to h/e^2 and h and e are the Planck constant and the electron charge, respectively.

Graphene, this single layer of carbon atoms in a hexagonal lattice, has remarkable electronic transport properties, which have been discovered only recently [77, 47]. Most of these properties are related to the chiral nature of the electrons in graphene. One the most famous is the unusual QHE [18], which follows a peculiar succession of quantized plateaus:

$$R_{\mathrm{H}} = R_{\mathrm{K}} \frac{1}{4\left(N+\frac{1}{2}\right)}. \tag{7.1}$$

This quantization sequence is specific to graphene and is often called half-integer QHE because of the 1/2 factor appearing in Eq. (7.1).

Bilayer and tri-layer graphene have also specific QHE with other specific sequences of plateaus. Also, the QHE requires that the material is of sufficient quality to be observed. The analysis of the QHE is, therefore, a powerful investigation method and the experimental observation of the QHE is of prime importance for most experimentalists struggling to obtain homogeneous graphene layers on various substrates.

Moreover, there are other motivations to study QHE in graphene. First, the validity of the theory of the QHE should be readily tested and investigated on such a new material. Second, the QHE relates a resistance to physical fundamental constants h and e, and its measurement can be remarkably precise and reproducible. In GaAs-based quantum wells, in which the QHE has been also observed for decades now, the values of the quantum Hall resistances agree to within an uncertainty of only a few hundreds of billion from one device to another [63]. This precision is so good that these systems are currently used in metrology institutes as quantum electrical resistance standards.

Remarkably, the QHE is expected to persist in graphene at higher temperatures and lower magnetic fields than in semiconductors. This property could open the way toward more convenient resistance standards based on graphene [54, 55] operating in relaxed experimental conditions compatible with compact helium-free cryomagnetic setups, which are more practical for the dissemination of the standard than the setup required by GaAs-based devices. In this chapter, we briefly review the different attempts of the last decade to obtain homogeneous large graphene flakes on silicon carbide (SiC). A special attention is given to QHE, as not only it can serve as a convincing proof of the good quality of graphene layer, but also as the last experimental results demonstrate that graphene can seriously challenge semiconductor for the realization of resistance standards. The main mechanisms limiting the mobility are also reviewed. Finally, other quantum effects affecting the conductivity are presented, and it is argued that they give information on both chiral nature of the charge carriers and type of disorder.

7.2 A Short Overview of Graphene

The graphene band structure has been known for more than 30 years. It was calculated in 1976 by Nagayoshi et al. [46]. The band structure can be determined by tight-binding calculation. Two important energy bands, referred to as the conduction band and the valence band, touch each other at the corners of the hexagonal first Brillouin zone. Two of these corners are non-equivalent. They define two valleys in the energy spectrum called K and K'. For undoped graphene, the valence band is full, the conduction band is empty, and the Fermi energy lies exactly at zero energy, at the K and K' points. Around the K and K' points, the tight-binding model gives, after additional simplification, the celebrated Dirac equation of graphene:

$$H_{\mathbf{K}} = v_{\mathbf{F}}\, \sigma \cdot p, \tag{7.2}$$

where $v_{\mathbf{F}}$ is the Fermi velocity ($v_{\mathbf{F}} = 10^6$ m/s), σ represents the Pauli matrices, p is the momentum operator, and $H_{\mathbf{K'}} = -H_{\mathbf{K}}$. The solution to the problem $H_{\mathbf{K}}|\Psi\rangle = E|\Psi\rangle$ is obtained by recognizing its formal equivalence to that of a real spin in a magnetic field. This pseudospin is aligned with the momentum and the wavefunctions have the form

$$|\psi \pm\rangle = \frac{1}{\sqrt{2}} \begin{pmatrix} e^{-i\theta_k/2} \\ \pm e^{i\theta_k/2} \end{pmatrix} e^{i\mathbf{k}\cdot\mathbf{r}}, \tag{7.3}$$

where $\theta_k = \arctan(k_y/k_x)$ and the $\pm$ signs correspond to the conduction and valence bands, respectively. The wavefunction changes sign upon the transformation $\theta_k \rightarrow \theta_k + 2\pi$. As the direction of propagation and the amplitude of the wave vector are not independent, the electrons are said to possess the property of chirality. The chiral nature of graphene has a strong influence on most of the transport properties, including QHE, but also the quantum corrections at low fields, namely weak localization and Altshuler–Aronov (AA) corrections.

The Landau level (LL) energies can be readily calculated by introducing a vector potential **A** in the Hamiltonian without magnetic field through the substitution $p = p + eA$. It follows that the energies of the LLs in graphene are given by:

$$E_{\mathrm{N}} = \pm v_{\mathrm{F}} \sqrt{2\hbar eBN}. \tag{7.4}$$

Table 7.1 sums up the main features of the physics of graphene and GaAs in the presence of magnetic field. In monolayer graphene, the energy of LLs varies as $\sqrt{B}$ instead of B in semiconductors. Also the energy gap between LLs reduces with the index number N. The degeneracy of each LL is $4eB/h$ (two spin directions and two valleys around the K and K' points). In the presence of disorder, the LL degeneracy is lifted. All states become localized, at the exception of one state per LL, which remains delocalized. The LL is still spin and valley degenerate and stays (approximately) at the same energy E_{N}. The QHE is a consequence of both LL formation and disorder. Laughlin [39] showed that when N delocalized states below the Fermi level are completely occupied, the Hall conductance is Ne^2/h and its value is independent of the position of the Fermi level as long as it lies in the localized states. The anomalous half-integer sequence of Hall plateau is related to the fact that the zero energy LL is half-filled for zero carrier density $n = 0$.

Numerical values indicated in Table 7.1 and Fig. 7.1 show that the energy spacing between first LLs in monolayer graphene is much larger than in GaAs at low magnetic induction. This explains the observation of the QHE quantization at 300 K [48] and generally speaking suggests that the Hall quantization in graphene should be very robust.

Table 7.1 Comparison of the LL parameters in GaAs and graphene

System	GaAs	Graphene
Degeneracy	$2eB/h$	$4eB/h$
Hall resistances	$(2N)^{-1}R_{\mathrm{K}}$	$\pm(4N + 1/2)-1R_{\mathrm{K}}$
Energy spacing	$\simeq 20\ B[T](\mathrm{K})$	$\simeq 420\sqrt{B[T]}\,(K)$

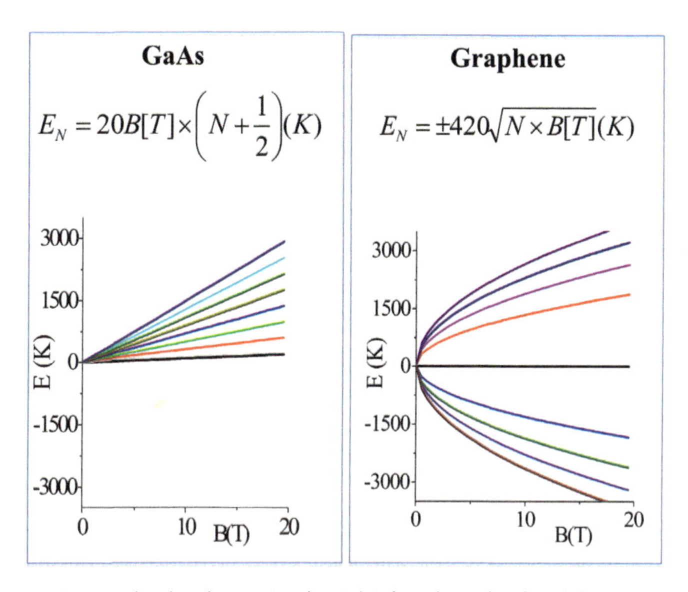

Figure 7.1 Landau level energies (in Kelvin) and Landau level dispersion as a function of magnetic field in GaAs-based 2DEG (left panel) and in graphene. (right panel). The different colors refer to different indices *N*. Black lines refer to *N* = 0. Red lines refer to *N* = 1.

7.3 Growth and Characterization of Graphene on SiC

The formation of graphite on the surface on heated SiC was first observed by van Bommel et al. [72] in 1975 by heating a crystal of silicon carbide around 1000°C. Epitaxial graphene forms by the sublimation of silicon atoms and the reorganization of the carbon atoms at the SiC surface. In 2004, Berger et al. observed Shubnikov–de Haas oscillations in a device made of a few layers of epitaxial graphene on the silicon face of SiC [6]. However, obtaining epitaxial graphene as good as exfoliated graphene [47, 77] revealed to be difficult and the observation of the QHE was belated.

Remarkable progresses were made when it was realized that growth should be done under a controlled atmosphere (e.g., under argon and not under high vacuum). Using this technique, several groups reported unambiguously the observation of the half-integer

QHE in epitaxial graphene before 2010. In 2009, Wu et al. observed the QHE on a single graphene layer grown on the carbon face (C-face) of 4H-SiC [76] (the notations 4H-SiC and 6H-SiC refer to the two most common polytypes of SiC). The mobility was 20,000 $cm^2/V\,s$ at 4 K and 15,000 $cm^2/V\,s$ at 300 K. They used the RF furnace method, in which the substrate is enclosed in a graphitic chamber with or without inert gas to produce high-quality graphene of arbitrary thickness. Camara et al. [7] reported the observation of the half-integer QHE in single-layer epitaxial graphene on the C-face of SiC prepared by argon-assisted graphitization. Large graphene ribbons (up to 50 μm large, hundreds of μm long) were obtained by this method. Electrical measurements revealed that graphene is p-type with a carrier concentration of 10^{12} cm^{-2} and mobility up to 10,000 $cm^2/V\,s$ at 4 K.

Despite these initial successes, graphene grown on the C-face of SiC was deceiving in the following years. On this face, graphene suffers from the lack of spatial homogeneity, a problem which has not been solved yet. It is currently not possible to obtain large monolayer flakes of several squared centimeters, as initially envisioned. The C-face is currently mostly used for the growth of multilayer graphene, which has also important electrical properties, in particular for spintronics [13, 20]. We will not focus here on these applications.

In the same period, significant progress was also made on the silicon face (Si-face). Epitaxial graphene grown on the Si-face of SiC has almost always been observed to be strongly n-type doped with carrier densities in the range of 10^{12}–10^{13} cm^{-2}. The doping of the graphene is caused by the so-called buffer-layer or zero-layer graphene (ZLG) of carbon atoms between the SiC substrate and graphene [14, 61, 73]. This ZLG is characterized by a large supercell $6\sqrt{3}\times 6\sqrt{3}$ of the reconstructed surface of sublimated SiC. Because of missing or substituted carbon atoms in the supercell [59], because of strong substrate–carbon bonds [73], the buffer layer does not inherit the properties of graphene. The ZLG band structure is made of localized surface states with a broad distribution of energies. The ZLG is non-conducting [59, 60], does not induce parasitic parallel conduction but strongly n-dopes the upper graphene layer. With dopings of 10^{12} cm^{-2} or more, the upper LLs $N \geq 1$ remain populated for magnetic fields below 20 T. Thus, the most robust and physically interesting quantum Hall plateau, which corresponds to the filling of the $N = 0$ LL only, can only be observed in very high magnetic fields. It is, therefore, crucial to compensate the doping induced by the ZLG, either by an appropriate gating of the structure, or by a modification of the interface.

Tanabe et al. (NTT Basic research laboratories) [67] grew high-quality monolayer graphene on the Si-face of SiC by thermal decomposition. The SiC substrates were annealed at around 1800°C in argon pressure of less than 100 torrs. The electrical properties were investigated in top-gated devices. To form the dielectric layer of the top gate, 85 nm of hydrogen silsesquioxane (HSQ) and 40 nm of SiO_2 were, respectively, spin-coated and sputtered onto the whole substrate. At 2 K, the carrier mobility of the graphene exceeded 10,000 $cm^2/V\,s$, and the half-integer QHE was observed at various carrier densities when top gate bias was applied. Despite the fact that the graphene channel was covered by the dielectric layer, the carrier mobility was relatively high, indicating the good intrinsic quality of graphene.

Pan et al. (Sandia National Laboratories, USA) [51] reported the observation of the integer quantum Hall states at LL filling factors ν = 2, 6, 10 in a Hall bar device made of a single-layer epitaxial graphene film on the Si-face of SiC prepared via argon-assisted graphitization. The LL filling factor ν is defined as $\nu = (eB/h)$ where n is the carrier density. The two-dimensional electron gas exhibited at 4 K a carrier mobility of 14,000 $cm^2/V\,s$ at the electron density of $6.1 \times 10^{11}\ cm^{-2}$.

Back-gate technology on Si-face epitaxial graphene devices was first demonstrated by Waldmann et al. [75]. They induced a conducting layer below the surface of the SiC crystal through implantation of nitrogen ions before the graphene growth. At low implantation dose, the implanted layer acted as a back gate and the carrier density could be changed through the Dirac point if the graphene layer was decoupled from the buffer layer through the intercalation of hydrogen. Without hydrogen intercalation, the ZLG states screened the gate potential and the gate was ineffective. A similar back-gate technique was used by Jouault et al. to tune the carrier density of a monolayer graphene on the carbon face of SiC [31]. No hydrogen passivation was required as there is no ZLG on this face. The half-integer QHE of graphene was demonstrated.

Jobst et al. used chemical gating by covering graphene with tetrafluorotetracyanoquinodimethane (F4-TCNQ) molecules. The carrier density was reduced to 5.4×10^{11} cm^{-2} with a mobility of 29,000 cm^2/V s and they demonstrated QHE.

Another possibility to control the carrier concentration is to modify the graphene–SiC interface. Pallecchi et al. demonstrated that oxygen adsorption reduces the carrier density in epitaxial graphene. The QHE was observed with a $\nu = 2$ plateau starting around 10 T [50]. However, the mobility was very low. Recently, the same group demonstrated that using post-growth annealing under hydrogen [49], the mobility can increase from 3000 cm^2/V s to more than 11,000 cm^2/V s at 0.3 K, whereas the electron concentration decreases to less than 10^{12} cm^{-2}. Alternatively, it was demonstrated that graphene can be doped by electrostatic potential gating with ions produced by corona discharge [38]. This method is currently one of the most versatile and efficient to reach very low carrier densities.

Lara-Avila et al. demonstrated non-volatile control of the carrier density by placing on top of graphene a layer of PMMA/MMA followed by another layer of ZEP520A, chosen for its ability to provide potent acceptors under deep UV light [37]. Using this technique, Tzalenchuk et al. were the first to demonstrate the viability of a quantum Hall resistance standard based on large-area epitaxial graphene synthesized on the Si-face of SiC. They demonstrated the agreement of the Hall resistance with $R_K/2$ with a relative standard measurement uncertainty down to 8.7×10^{-11} at $B = 14$ T and $T = 0.3$ K [25], and slightly lower than 10^{-9} at $B = 11.5$ T and $T = 1.5$ K [28]. The agreement was shown within a few parts per billion at $T = 1.3$ K and $B = 14$ T [71]. Graphene was produced by the Linköping University. It was grown at 2000°C and 1 atm argon gas pressure, resulting in monolayers of graphene atomically uniform over more than 50 μm^2. The manufactured Hall bar was n-doped, owing to charge transfer from SiC, with the measured electron concentration lying in the range 5×10^{11} to 10×10^{-11} cm^{-2}, mobility of 2400 cm^2/V s at room temperature and between 4000 and 7500 cm^2/V s at 4.2 K, almost independent of device dimensions and orientation with respect to the substrate terraces.

Another group also used samples produced by Graphensic AB, a spin-off from the Linköping University. After the lithography process, low carrier densities were obtained and the Hall resistance was measured on the $\nu = 2$ plateau. The accuracy of the quantized Hall resistance was not demonstrated with a relative uncertainty better than a few 10^{-7} at 3 T and approximately 1.8×10^{-9} at 8 T [62]. Moreover, a large dispersion of the measurements (up to 0.5×10^{-6} in relative value) was observed by changing the Hall terminal pairs used, manifesting strong inhomogeneity in these very large samples.

A modification of the growth on the Si-face of SiC was initiated by Michon et al. [43]. These authors obtained monolayer graphene, but not by a pure sublimation method. The monolayer graphene film was grown by propane/hydrogen CVD on SiC [24], a hybrid technique that allows the tuning of the electronic transport properties [21, 23, 24, 57, 66].

A probable advantage of CVD growth resides in its versatility. CVD allows graphene to grow either on the standard ZLG, or on a hydrogenated interface (nicknamed quasi-freestanding graphene) [65]. The nature of the interface depends closely on the growth conditions and one of the key parameters is the growth temperature. Whatever the interface, the graphene grown by CVD appears extremely homogeneous over several centimeters under optical investigations.

Graphene films obtained by CVD at high temperatures appeared similar to those obtained by the sublimation method. In both cases, the surface is altered by the formation of SiC steps during the growth. Also both processes lead to the formation of carbon-rich interface layer (ZLG), which strongly n-dopes the graphene monolayer. On the contrary, graphene films obtained at lower temperatures appear similar to monolayer graphene films with post-growth hydrogenated interface [60]. They have a slightly higher mobility, a p-type doping and are strongly modified by a temperature elevation above 700°C.

Jabakhanji et al. performed measurements on graphene on a hydrogenated interface at very high magnetic fields (up to 58 T) and observed the half-integer QHE specific to graphene up to $\nu = 4$ [24] (see Fig. 7.2). This graphene was intrinsically highly p-doped, preventing the observation of the quantum Hall plateau at $\nu = 2$ and accurate measurements of the Hall quantization were not possible because so high magnetic fields are obtained in short pulses only. The origin of this doping is believed to be extrinsic (defects, resist).

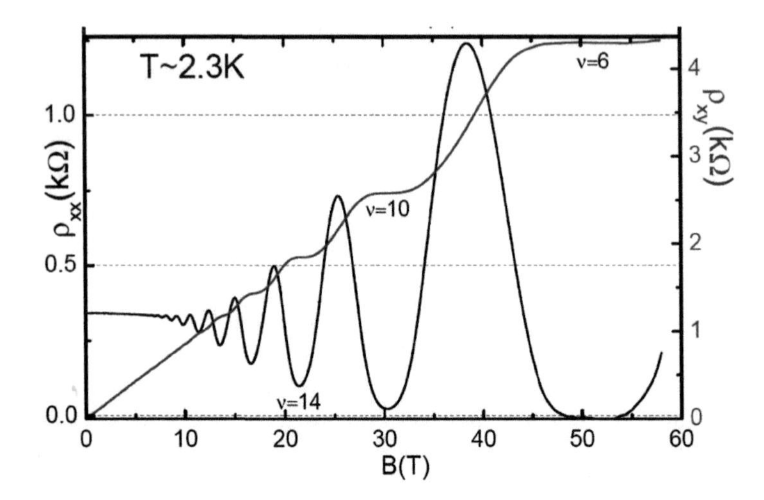

Figure 7.2 Magnetoresistances at low temperature (T = 2.3 K) up to B = 58 T, revealing the half-integer quantum Hall effect in a graphene monolayer grown by CVD on SiC. The ZLG interface has been passivated by hydrogen during the growth. The hydrogenation leads to a dramatic change in the interface and a high p-doping results. The measurements have been performed with a Hall bar of width 100 μm and length 500 μm. Reprinted from Ref. [24], with permission from American Physical Society.

Lafont et al. used CVD graphene on the Si-face of SiC, which resides on the carbon-rich ZLG interface. This graphene is highly n-doped after the growth. However, this doping was reduced by post-growth annealing and by encapsulating the graphene under a resist bilayer, as done previously by Lara-Avila et al. [37].

Even without intentional illumination, the final concentration, around 3.2×10^{11} cm^{-2}, allowed the observation of the $\nu = 2$ quantum plateau and accurate measurements of the QHE (see Fig. 7.3). Lafont et al. found an agreement of the Hall resistance with $R_K/2$ within a relative measurement uncertainty of 1×10^{-9} at T = 1.4 K in a magnetic field range between 10 T and 19 T. By averaging measurements, performed at several magnetic field values, an overall accuracy of $(-2 \pm 4) \times 10^{-10}$ was also determined. These Hall quantization measurements showed that graphene can operate accurately at magnetic fields as low as the reference standards based on GaAs, at a similar temperature of 1.4 K. This was, therefore, the first demonstration that a graphene device can directly replace a GaAs-based quantum resistance standard in a conventional setup of

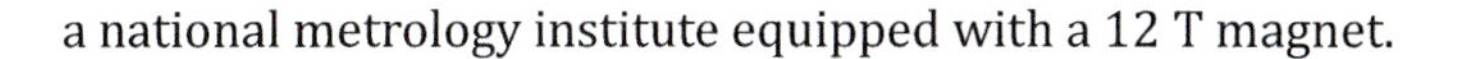

a national metrology institute equipped with a 12 T magnet.

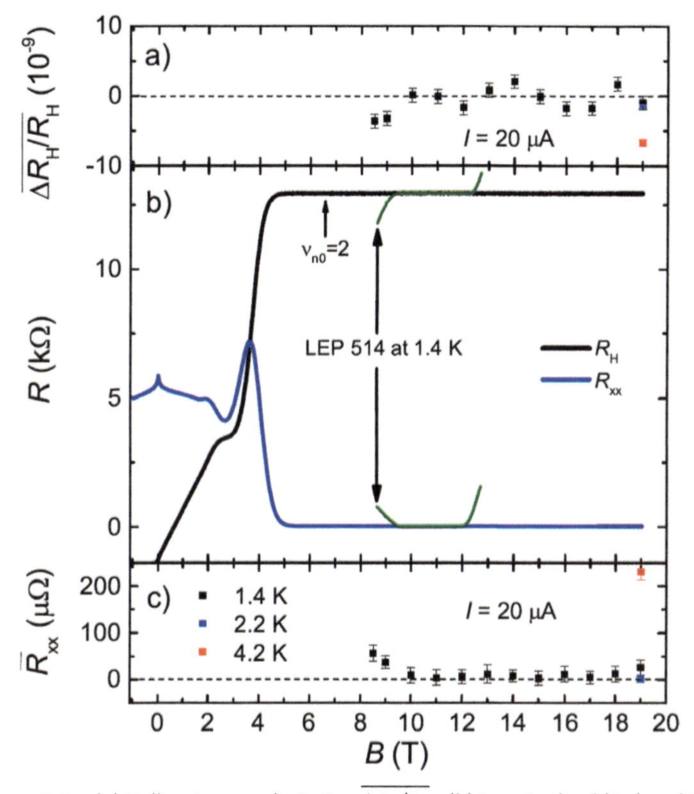

Figure 7.3 (a) Hall resistance deviation $\overline{\Delta R_H/R_H}$. (b) Longitudinal ($R_{xx}$) and Hall ($R_{xy}$) resistances for B varying from 0 T to 19 T for the graphene sample (blue curve) and varying from 8 T to 13 T for the GaAs sample (green curve). ν_{n0} is the Landau level filling factor calculated from the carrier density n_0 determined at low magnetic fields. (c) Accurate measurements of the longitudinal resistance $\overline{R_{xx}}$ versus B. Error bars correspond to one standard deviation [36].

7.4 Transport Properties of Graphene on SiC

7.4.1 Evidences of Hopping Conduction in the Quantum Hall Regime

The study of the resistivity as a function of temperature is an important source of information on disorder. At sufficiently low temperatures, in semiconductor-based two-dimensional systems, the longitudinal resistivity in the quantum Hall regime is governed by what is known as variable range hopping (VRH). In this situation, the Fermi level lies between localized states far from the centers of the LLs. However, the conductivity is nonzero because the electrons may hop from one localized state to another one, assisted by phonons. At more elevated temperature, because of the smearing of the Fermi function, the extended states can be thermally populated and the resistivity then simply scales as $\exp(\Delta E/k_B T)$, where the activation energy ΔE is of the order of the energy separation between adjacent LLs.

In semiconductor-based heterostructures, like GaAs quantum wells, the transition between the two regimes is typically of the order of 4 K [15]. In graphene, the transition occurs at a much more elevated temperature of the order of 100 K, because the energy separation between LLs is much larger (of the order of 1000 K between the lowest LLs for a few teslas).

In most experimental cases, below $T \simeq 100$ K, the conductivity in the quantum Hall regime of graphene is driven by the Efros–Shklovskii VRH. This conclusion is valid for both exfoliated [5, 17] and graphene on SiC [26]. Some of the most recent and convincing measurements in graphene on SiC have been reported by Lafont et al. Figure 7.4a reproduces $\sigma_{xx}(T) \times T$ plotted in logarithmic scale as a function of $T^{-1/2}$, where $\sigma_{xx} = R_{xx}/\left(R_{xx}^2 + R_{xy}^2\right)$ is the longitudinal conductivity. The linearity of the curves over five orders of magnitude fits precisely the description $\sigma_{xx}(T) \times T = \sigma_0(B) \exp(-(T_0(B)/T)^{1/2})$ where $T_0(B)$ and $\sigma_0(B)$ are B-dependent fitting parameters, as expected from a dissipation mechanism based on VRH with a soft Coulomb gap [64]. As expected, thermal activation does not fit the data and does not manifest itself in the investigated temperature range, below 100 K.

It is possible theoretically to link the VRH law to the localization length $\xi(B)$, which measures the spatial extension of the localized states participating to the conduction. The relation is $k_B T_0(B) = Ce^2/4\pi\varepsilon_r\varepsilon_0\xi(B)$ where $C = 6.2$ [40, 15], k_B is the Boltzmann constant, ε_0 is the permittivity of free space, ε_r is the mean relative permittivity of the graphene on SiC covered by the resists. Figure 7.4b shows the evolution of both $T_0(B)$ and $\xi(B)$. It appears first that $\xi(B)$ continuously decreases from 10.5 nm to 5.5 nm between $B = 7$ T and $B = 19$ T. By contrast, in GaAs-based devices, well-defined minima of $\xi(B)$ are observed at magnetic fields corresponding to integer values of the filling factor [15]. This is not the case for graphene (see Fig. 7.4b). The decrease in $\xi(B)$ suggests that a charge transfer takes place between the interface states and graphene, as proposed by Kopilov et al. [35] and previously experimentally observed by Janssen et al. [27]. This transfer leads to a constant filling factor value over a magnetic field range of a few teslas.

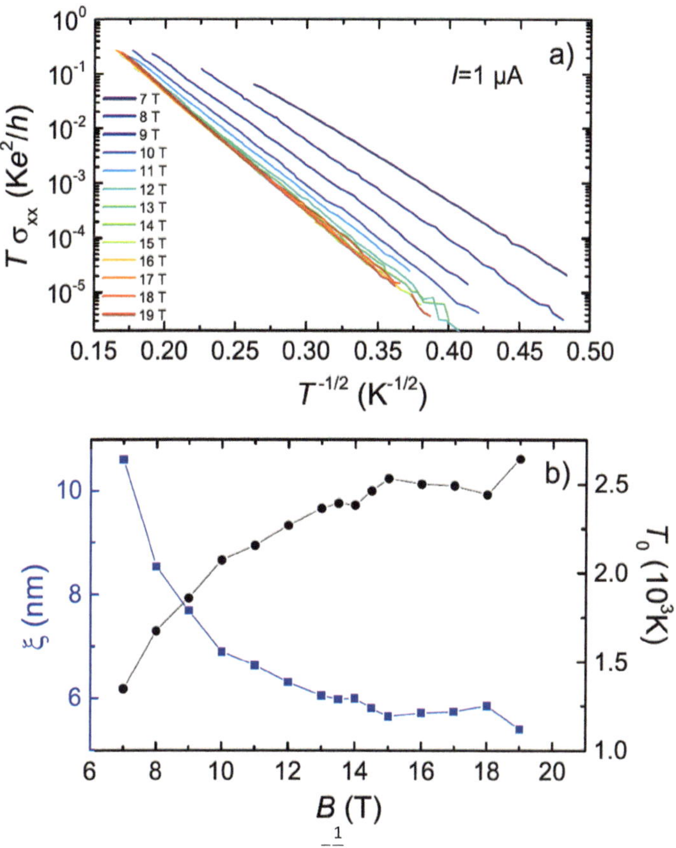

Figure 7.4 (a) $T\sigma_{xx}$ as a function of $T^{-\frac{1}{2}}$ in a semi-log scale for magnetic fields from 7 T to 19 T and in a temperature range from 4 K to 40 K. (b) Temperature parameter T_0 (black circles, right axis) and localization length ξ (blue squares, left axis) obtained from the adjustment of curves of figure (a) by the VRH model as a function of the magnetic field. The dissipation in the QHE regime is well described by a VRH mechanism with a soft Coulomb gap. The continuous decrease in the localization length as the magnetic field increases, without showing a minimal value, explains the robustness of the Hall resistance plateau toward high magnetic fields [36].

Even more remarkable, $\xi(B)$ is extremely small in graphene and is comparable to the cyclotron radius. This suggests that ξ is imposed by some specific disorder, of unknown origin, which leads to an extreme localization. This localization increases the robustness of the QHE with respect to the temperature.

7.5 Scattering Mechanisms Limiting the Mobility

Despite the remarkable observation of the half-integer QHE in graphene on SiC, the mobility remains modest. By contrast, it is established that the intrinsic mobility of graphene is limited by phonons. This intrinsic mobility is estimated to be around 200,000 $cm^2/V\,s$ at room temperature [45]. It is limited by flexural phonons and has a strong $1/T^2$ dependence [8]. Why is this mobility of graphene on SiC so low, and how to increase it?

7.5.1 Main Scattering Mechanisms on Graphene on SiO_2

The understanding of the electronic properties of graphene on SiC is enlightened by the enormous amount of experimental and theoretical works that have been performed for exfoliated graphene on SiO_2. In this case, the experimental facts are a relatively fair mobility (around 10,000 $cm^2/V\,s$ at room temperature) and a weak temperature dependence of conductivity. This means that for graphene on SiO_2, the intrinsic temperature-dependent contribution to the resistivity is negligible in comparison with the extrinsic contribution, e.g., scattering by defects.

For charged (Coulomb) impurities, theoretical quantitative estimations give [1]:

$$\sigma(n) \simeq K(e^2/h)(n/n_{\mathrm{imp}}), \tag{7.5}$$

where n_{imp} is the concentration of charged impurities and K is a prefactor that depends on the screening properties of the environment. For graphene on SiO_2, $K \simeq 20$. For graphene on SiC, $K \simeq 30$. The dependence of the conductivity σ on the charge carrier concentration n is often linear for graphene on SiO_2 [47], while the mobility is weakly dependent on the concentration. Early experiments, therefore, concluded that charged impurities control the mobility. However, further experimental works found that in high-quality samples, the mobility is too weakly dependent on the dielectric constant of the environment to be explained by screened Coulomb interaction only [56]. Other scattering mechanisms must be at play.

With charged impurities, resonant scattering mechanism is currently considered one of the dominant processes limiting the electronic mobility in graphene. This mechanism is induced by very short-range potentials, like adsorbed atoms or molecules that bind to one of the carbons of graphene to form σ-bond. This type of defects [53] gives a sublinear contribution to the conductivity:

$$\sigma = 2\frac{e^2}{\pi h}\frac{n}{n_i}\ln^2(R\sqrt{\pi n}), \tag{7.6}$$

where n_i and R are the concentration and the size of the short-range scattering potentials. By combining Coulomb scattering with short-range resonant scattering, most of the experimental results can be reproduced [44, 53].

7.5.2 Graphene on SiC: Short-Range Defects, Interface Phonons, and Structural Steps

This overall picture must be mended for graphene on SiC. First of all, the mobility is rather low. The mobility at 300 K is around 1000 cm^2/V s with weak temperature dependence. Therefore, the conductivity is obviously determined by extrinsic mechanisms.

Moreover, in an early experimental work, Tedesco et al. [69] reported a constant conductivity *versus* n on the Si-face and a sublinear dependence of σ *versus* n on the C-face, which implies that the mobility depends on n. Thereafter, this dependence was confirmed by several works [22, 31, 68, 75]. The dependence is evidenced in Fig. 7.5, which reproduces the results of Tanabe et al. [68]. At very low doping, Tanabe et al. also measured a record mobility of 45,000 cm^2/V s at T = 2 K. These results suggest that short-range defects/resonant scatterers can play a role. However, the mobility also decreases with increasing temperatures, a strong indication that beyond resonant scattering other mechanisms are at work.

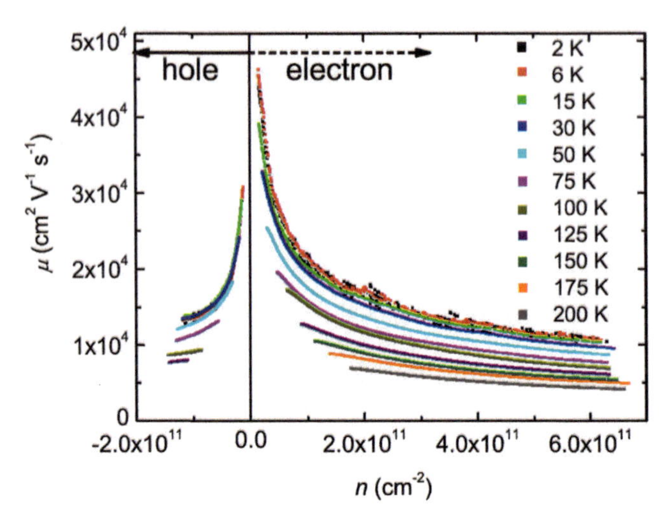

Figure 7.5 Hole and electron mobility as a function of carrier density and temperature for a graphene layer on the Si-face of SiC. Reprinted from Ref. [68], with permission from American Physical Society and kind courtesy of Prof. Hiroki Hibino, Kwansei Gakuin University, Japan.

Besides, the role of the interface was enlightened in 2011 by the work of Speck et al. [65], who observed an improvement in the mobility by using hydrogen intercalation, a technique that replaces the defective carbon-rich ZLG layer by a hydrogen-passivated interface. Lots of further experimental works confirmed that the presence of the ZLG interface is extremely detrimental to the mobility. Recently, interface acoustic phonons were identified as a dominant source of scattering. Low-energy phonon modes in graphene on SiC were identified by inelastic tunneling spectroscopy (IETS) [9]. Two modes were found at E_1 = 70 meV and E_2 = 16 meV. The presence of these modes was shown to depend strongly on the presence of defects in the buffer layer. These phonon modes are quite close to the out-of-plane acoustic phonon modes at 16 and 58 meV measured in graphite [74]. Several groups successfully modeled the temperature dependence of the mobility by taking into account the activated conductivity induced by similar low-energy phonon modes [11, 16, 23, 68]. Giesbers et al. observed a relation $\mu \propto 1/\sqrt{n}$ in a few samples of good quality [16]. This dependence is the one predicted for scattering on surface polar phonons on polar substrates [52].

Finally, another important factor to improve the mobility is reducing the influence of the surface steps, which are almost inevitably present on the SiC substrate. Even nominally on-axis substrates are not step-free after graphene growth. The in-plane anisotropy of carrier mobility of graphene on SiC with various off-angle of the substrate was reported by various groups [12, 33]. Actually, a graphene device patterned perpendicularly to the surface steps can have a resistance higher by one order of magnitude than a similar device patterned parallel to the steps. The specific effect of a single step was investigated by Low et al. [41]. These authors demonstrated that the resistivity of graphene covering the step depends on the step height. This indicates that the dominant scattering mechanism is due to inhomogeneity in the carrier density, which is reduced on the lower side of the step edge where graphene is pulled away from the substrate. The detrimental role of the steps on the mobility is also increased by the frequent presence of multilayer patches at the step edges. Because the transmission from monolayer to bilayer is not perfect, these patches induce additional resistances at the step edges [29].

7.6 Quantum Corrections at Low Fields

The QHE is not the only manifestation of quantum phenomena affecting the conductivity. At low temperature and low magnetic field, the Drude model, or even the semiclassical Boltzmann equation, is not enough to describe the properties of two-dimensional electron gases, and quantum corrections have to be taken into account. These corrections are the result of quantum interference of electrons moving along different trajectories allowed by impurity scattering. There exist two quantum corrections to the conductivity called weak localization correction, which results from interference of time-reversed trajectories, and Altshuler–Aronov (AA) correction, which is induced by electron–electron interaction only. The quantum corrections can be experimentally detected by their characteristic dependence on magnetic field and temperature. Their measurement provides important information about the type of disorder, the conservation of the chiral nature of the particles, and the strength of the particle interaction.

7.6.1 Weak Localization

It appears that the physics of quantum corrections in graphene is extremely complex because of the chiral nature of the particles, which prevents these particles to be backscattered. For the sake of simplicity [53], let us assume a scattering potential to have the form

$$V(r) \propto \exp\left(-r^2/r_0^2\right), \tag{7.7}$$

where r_0 defines the range of the potential. One can calculate the effect of this potential on electrons in the same valley:

$$\left\langle \Psi_+(\boldsymbol{k}') | V | \Psi_+(\boldsymbol{k}) \right\rangle \propto f(\boldsymbol{k},\boldsymbol{k}')\exp(-q^2 r_0^2/4), \tag{7.8}$$

where $f(k, -k') = \cos(\theta_k/2 - \theta_{k'}/2)$ and $q = |k - k'|$. It appears that $f(k, -k) = 0$ and thus backscattering is forbidden because of the Berry phase π. However, the situation is different if we calculate the scattering between electrons from different valleys. It is then given by:

$$\left\langle \Psi_+(\boldsymbol{k}') | V | \Psi_+(\boldsymbol{k}) \right\rangle \propto g(\boldsymbol{k},\boldsymbol{k}')\exp\left(-Q^2 r_0^2/4\right), \tag{7.9}$$

where $g(k, k') = \sin(\theta_k/2 - \theta_{k'}/2)$ and $Q = |K - K'| = 4\pi/(3\sqrt{3a})$, $a \simeq 1.4$ Å is the distance between adjacent carbon atoms. As $g(k, k') \neq 0$, backscattering is allowed. Moreover, only very short-range potentials with $r_0 < a$ have significant scattering amplitude.

Thus, in the absence of short-range potentials, backscattering remains forbidden. This leads to weak antilocalization (increase in conductivity) in graphene instead of the usual weak localization (decrease in conductivity) observed in most two-dimensional systems without spin–orbit scattering. This picture starts to break down at high energies ($\simeq$1 eV above the Dirac point), where the circular symmetry around the K and K' points is deformed by the trigonal symmetry of the graphene lattice, which breaks the interference of time-reversal symmetric paths and decreases the weak antilocalization correction. Finally, in the presence of short-range scattering, intervalley scattering restores backscattering and weak localization replaces weak antilocalization.

Actually, the magnetic field dependence for the weak (anti) localization in graphene has been calculated for the diffusive regime [2, 42]. We reproduce this result, which is often used by experimentalists in the literature:

$$\Delta\rho(B) = -\frac{e^2\rho^2}{\pi}\left[F\left(\frac{\tau_B^{-1}}{\tau_\varphi^{-1}}\right) - F\left(\frac{\tau_B^{-1}}{\tau_\varphi^{-1} + 2\tau_i^{-1}}\right) - 2F\left(\frac{\tau_B^{-1}}{\tau_\varphi^{-1} + \tau_i^{-1} + \tau_*^{-1}}\right)\right], \tag{7.10}$$

where $F(z)$ = ln$z$ + $\Psi(1/2 + 1/z)$, Ψ is the digamma function, $\tau_B^{-1} = 4eDB/\hbar$, and D is the diffusion constant. The various scattering times taking place in this equation are the coherence time τ_φ, the intervalley scattering time τ_i, and τ_*, which regroups several scattering times: $\tau_*^{-1} = \tau_w^{-1} + \tau_z^{-1}$, where τ_w is the elastic intravalley trigonal warping scattering time and τ_z itself regroups all other elastic intravalley chirality breaking scattering times.

Hopefully, this equation can be simplified at small magnetic fields, where $F(z) \simeq z^2/24$, and the correction can be written as:

$$\Delta\rho(B) = -\frac{e^2\rho^2}{\pi}\left(\frac{4eDB\tau_\varphi}{\hbar}\right)^2\left[1 - \frac{1}{\left(1 + \frac{2\tau_\varphi}{\tau_i}\right)^2} - \frac{2}{\left(1 + \frac{\tau_\varphi}{\tau_*} + \frac{\tau_\varphi}{\tau_i}\right)^2}\right]. \tag{7.11}$$

This shows that the amplitude of the weak localization correction is governed by τ_φ, whereas the sign of the correction depends on the ratio of the scattering times. The favorable conditions for the observation of positive magnetoresistance correspond to small τ_φ/τ_* and τ_φ/τ_i.

Experimentally, it appears that weak localization (negative magnetoresistance at low fields) has been observed in almost all graphene samples studied so far, whereas weak antilocalization has been very rarely observed. The only unambiguous observation of weak antilocalization has been reported by Tikhonenko et al. [70], by studying the gate and temperature dependence of the quantum corrections in exfoliated graphene. In graphene on SiC, weak localization has been, up to now, systematically observed at zero magnetic field. As the magnetic field increases, the weak localization peak fades out and a positive magnetoresistance is often observed and interpreted as a signature of weak antilocalization. This interpretation is questionable, because in graphene on SiC, the mobility is low and the range of validity of Eq. 7.12 overlaps with the magnetic field range on which geometric corrections are already sizeable. Geometric corrections lead to a parabolic positive magnetoresistance, which is experimentally difficult to distinguish from the weak antilocalization correction.

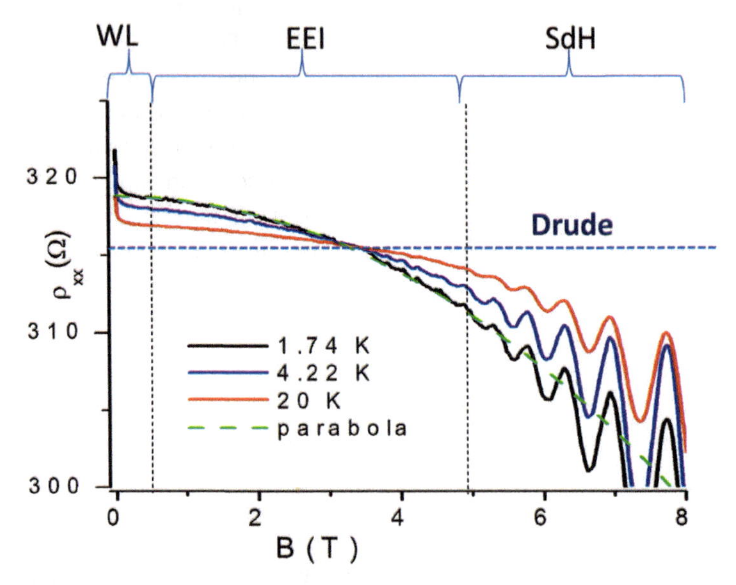

Figure 7.6 Magnetoresistivity of a monolayer graphene Hall bar, evidencing the weak localization peak at $B < 0.5$ T, the Altshuler–Aronov correction (EEI) in the range $B = 0.5$–5 T and the Shubnikov–de Haas oscillations at higher fields. The electron concentration is around $n = 10^{13}$ cm^{-2}. Reprinted by Ref. [23], with permission from American Physical Society.

Whatever the case, the presence of a weak localization peak indicates that intervalley scattering is sizeable in all samples of graphene on SiC currently realized. Intervalley scattering is directly linked to the presence of very short-range potentials, such as adatoms, adsorbed hydrocarbon, and vacancies. The nature of this disorder probably varies from sample to sample. The typical intervalley scattering time is of the order of $\tau_i \simeq 1$–10 ps, and the typical intervalley scattering length $\sqrt{D\tau_i}$ varies in the range 100–200 nm [4, 22, 30, 32]. Finally, the intervalley scattering time is systematically lower than the transport time extracted from the mobility, indicating that other sources of scattering are also present.

7.6.2 Altshuler–Aronov Correction

Even if a weak magnetic field is applied, just enough to kill the weak localization effect, the Drude conductivity is not retrieved yet. This is because one has to consider scattering of interacting quasi-particles at impurities. Back in 1979, Altshuler and Aronov [3] predicted that the conductivity of such a system acquires an additional temperature dependence of the type similar to the one coming from weak localization:

$$\delta\sigma = -\frac{e^2}{2\pi^2\hbar} A\left(F_0^\sigma\right) \ln\frac{\hbar}{k_B T \tau_{tr}}, \tag{7.12}$$

which is valid when $k_B T \tau_{tr}/\hbar \ll 1$ and τ_{tr} is the transport time, also named momentum relaxation time. The coefficient A depends on the Fermi liquid constant F_0^σ, an important parameter in the theory of quantum liquids, which measures the strength of the spin-exchange interaction. If $F_0^\sigma < 0$, the spin-exchange interaction tends to align electron spins. For weakly interacting electron gas, A is positive and $\simeq 1$.

By a simple matrix inversion, the AA correction $\Delta\rho$ to the resistivity ρ is obtained from Eq. (7.12):

$$\Delta\rho = -A\left[\mu^2 B^2 - 1\right] \frac{e^2\rho^2}{2\pi^2\hbar} \ln\left(\frac{\hbar}{k_B T\, \tau_{tr}}\right) \tag{7.13}$$

and its characteristic magnetic field dependence makes this formula especially useful for experimentalists. In graphene on SiC, due to the large graphene surfaces and the overall good sample homogeneity, the negative parabolic magnetoresistance induced by the last formula has been reported by several groups. All magnetoresistance curves, taken at sufficiently low temperatures, cross at $\mu B \simeq 1$, as expected from Eq. (7.13). From the coefficient A, extracted from the magnetoresistance curvature, the Fermi liquid constant can be determined. This constant is found to be very small and negative, with values varying between −0.16 and 0. This indicates that in graphene, the spin-exchange interaction is much weaker than in most metals and degenerate semiconductors. For comparison, in GaAs and Si quantum wells, the Fermi liquid constant is larger ($F_0^\sigma \simeq -0.2$) for a comparable $r_c \simeq 1$. (Here, r_c is the interaction parameter, defined as the ratio of the Coulomb energy to the Fermi energy.)

To explain this relatively low value of F_0^σ found in all experiments, it was emphasized that F_0^σ is lowered in graphene due to the chirality of charge carriers, which prevents backscattering and thus also reduces the electron–electron interaction. When the chirality is taken into account, numerical estimation gives $F_0^\sigma \simeq 0.09$, a value indeed quite close to most of the experimental estimations.

Besides, in graphene, the constant A does not depend only on F_0^σ but also depends on the interplay of the intravalley and intervalley scattering times. This dependence should be visible in the temperature evolution of A, which would, in turn, gives information on the nature of the disorder. A modification of the temperature dependence of A was indeed observed in graphene on SiC and attributed to this effect [30].

Finally, the Altshuler–Aronov theory was developed for the case of low temperatures $k_B T\tau_{tr}/\hbar \ll 1$, in a sense that electrons move diffusively during their interaction. It was only recently realized that this condition is sufficient but not necessary for the occurrence of the effect. In the opposite limit, $k_B T\tau_{tr}/\hbar \gg 1$, electrons interact with one impurity at a time; thus, the problem can be formulated as a ballistic scattering theory for interacting electrons. The temperature evolution of the resistivity is then expected to be linear in T and disorder dependent [10, 19]. Currently, too few experiments have been performed at high temperatures to understand if the AA correction can give reliable information on the type of disorder in graphene. Nevertheless, Jabakhanji et al. studied this ballistic regime and observed an important enhancement of the AA correction, which they attributed to the presence of additional short-range scatterers [23].

7.7 Conclusion

We have given a short overview of the electronic features of epitaxial graphene on SiC. Using the Si- and C-terminated surfaces, graphene grows in a wide variety of forms. In contrast to the other graphene growth techniques, graphene on SiC can provide wafer-scale homogeneous graphene directly on a semi-insulating substrate. The situation is especially promising on the Si-face, where monolayer graphene of a few inches large can be obtained and QHE was observed with a very good quantization. At the time of writing, two independent groups have already measured graphene quantum Hall resistance standard with accuracies within 1×10^{-9} or less. This 1×10^{-9} target corresponds to the maximal uncertainty accepted in national metrology institutes to validate a device as a good primary quantum resistance standard. In view of these results, obtained only 10 years after the first measurements of the exceptional electronic properties of graphene, it seems extremely likely that graphene will realize its promises and will outperform soon other well-studied semiconductors in the very demanding resistance metrology application.

We also briefly reviewed other quantum phenomena that can be observed on monolayer graphene on SiC. While weak antilocalization is predicted for graphene, all graphene on SiC samples present a localization peak at low magnetic field, which is experimentally very similar to what can be observed in disordered two-dimensional electron gases. This peak is a signature of strong intervalley scattering, whose origin can be traced back to local defects as adatoms or vacancies. Concerning AA corrections, the electron–electron interaction is extremely small even at temperatures as low as 1.5 K. The Dirac nature of the quasi-particles remains elusive here. Nevertheless, these quantum corrections can be studied in greater details in graphene on SiC than in graphene on SiO_2, because of the large graphene surface, which kills mesoscopic fluctuations and allows for larger currents. They give valuable information on the nature of the disorder, which complements the more usual analysis of the mobility.

Obviously, there is still plenty of space for technological progress, leaving plenty of opportunity for future investigation. Graphene on SiC is a material full of promises, not only in the metrology, as outlined in this paper, but also for electronic, spintronic, and optical applications.

References

1. Adam, S., Hwang, E. H., Galitski, V. M., and Das Sarma, S. (2007). A self-consistent theory for graphene transport, *Proc. Natl. Acad. Sci. U.S.A.*, **104**, 18392.

2. Aleiner, I. L. and Efetov, K. B. (2006). Effect of disorder on transport in graphene, *Phys. Rev. Lett.*, **97**, 236801.
3. Altshuler, B. and Aronov, A. G. (1985). Electron–electron interaction in disordered conductors, in *Electron–Electron Interactions in Disordered Systems*, Efros, A.L. and Pollak, M. (Eds.), Elsevier.
4. Baker, A. M. R., Alexander-Webber, J. A., Altebaeumer, T., Janssen, T. J. B. M., Tzalenchuk, A., Lara-Avila, S., Kubatkin, S., Yakimova, R., Lin, C.-T., Li, L.-J., and Nicholas, R. J. (2012). Weak localization scattering lengths in epitaxial, and CVD graphene, *Phys. Rev. B*, **86**, 235441.
5. Bennaceur, K., Jacques, P., Portier, F., Roche, P., and Glattli, D. C. (2012). Unveiling quantum Hall transport by Efros-Shklovskii to Mott variable-range hopping transition in graphene, *Phys. Rev. B*, **86**, 085433.
6. Berger, C., Song, Z., Li, T., Li, X., Ogbazghi, A. Y., Feng, R., Dai, Z., Marchenkov, A. N., Conrad, E. H., First, P. N., and de Heer, W. A. (2004). Ultrathin epitaxial graphite: A 2D electron gas properties and a route toward graphene-based nanoelectronics, *J. Phys. Chem. B*, **108**, 19912.
7. Camara, N., Jouault, B., Caboni, A., Jabakhanji, B., Desrat, W., Pausas, E., Consejo, C., Mestres, N., Godignon, P., and Camassel, J. (2010). Growth of monolayer graphene on 8° off-axis 4H-SiC(000-1) substrates with application to quantum transport devices, *Appl. Phys. Lett.*, **97**, 093107.
8. Castro, E. V., Ochoa, H., Katsnelson, M. I., Gorbachev, R. V., Elias, D. C., Novoselov, K. S., Geim, A. K., and Guinea, F. (2010). Limits on charge carrier mobility in suspended graphene due to flexural phonons, *Phys. Rev. Lett.*, **105**, 266601.
9. Cervenka, J., van de Ruit, K., and Flipse, C. F. J. (2010). Giant inelastic tunneling in epitaxial graphene mediated by localized states, *Phys. Rev. B*, **81**, 205403.
10. Cheianov, V. V. and Fal'ko, V. I. (2006). Friedel oscillations, impurity scattering, and temperature dependence of resistivity in graphene, *Phys. Rev. Lett.*, **97**, 226801.
11. Chen, J.-H., Jang, C., Xiao, S., Ishigami, M., and Fuhrer, M. S. (2008). Intrinsic and extrinsic performance limits of graphene devices on SiO_2, *Nat. Nanotech.*, **3**, 4, 206.
12. Dimitrakopoulos, C., Grill, A., McArdle, T. J., Liu, Z., Wisnieff, R., and Antoniadis, D. A. (2011). Effect of SiC wafer miscut angle on the morphology and Hall mobility of epitaxially grown graphene, *Appl. Phys. Lett.*, **98**, 222105.
13. Dlubak, B., Martin, M.-B., Deranlot, C., Servet, B., Xavier, S., Mattana, R., Sprinkle, M., Berger, C., De Heer, W. A., Petroff, F., Anane, A., Seneor,

P., and Fert, A. (2012). Highly efficient spin transport in epitaxial graphene on SiC, *Nat. Phys.*, **8**, 7, 557.

14. Emtsev, K. V., Speck, F., Th, Ley, L., and Riley, J. D. (2008). Interaction, growth, and ordering of epitaxial graphene on SiC(0001) surfaces: A comparative photoelectron spectroscopy study, *Phys. Rev. B*, **77**, 155303.
15. Furlan, M. (1998). Electronic transport and the localization length in the quantum Hall effect, *Phys. Rev. B*, **57**, 14818.
16. Giesbers, A. J. M., Procházka, P., and Flipse, C. F. J. (2013). Surface phonon scattering in epitaxial graphene on 6H-SiC, *Phys. Rev. B*, **87**, 195405.
17. Giesbers, A. J. M., Zeitler, U., Ponomarenko, L. A., Yang, R., Novoselov, K. S., Geim, A. K., and Maan, J. C. (2009). Scaling of the quantum Hall plateau-plateau transition in graphene, *Phys. Rev. B*, **80**, 241411.
18. Goerbig, M. O. (2011). Electronic properties of graphene in a strong magnetic field, *Rev. Mod. Phys.*, **83**, pp. 1193–1243.
19. Gornyi, I. V. and Mirlin, A. D. (2004). Interaction-induced magnetoresistance in a two-dimensional electron gas, *Phys. Rev. B*, **69**, 045313.
20. Han, W., Kawakami, R. K., Gmitra, M., and Fabian, J. (2014). Graphene spintronics, *Nat. Nanotech.*, **9**, 10, 794.
21. Hwang, J., Shields, V., Thomas, C., Shivaraman, S., Hao, D., Kim, M., Woll, A., Tompa, G., and Spencer, M. (2010). Direct growth of few-layer graphene on 6H-SiC and 3C-SiC/Si via propane chemical vapor deposition, *J. Cryst. Growth*, **312**, 3219.
22. Iagallo, A., Tanabe, S., Roddaro, S., Takamura, M., Hibino, H., and Heun, S. (2013). Tuning of quantum interference in top-gated graphene on SiC, *Phys. Rev. B*, **88**, 235406.
23. Jabakhanji, B., Kazazis, D., Desrat, W., Michon, A., Portail, M., and Jouault, B. (2014a). Magnetoresistance of disordered graphene: From low to high temperatures, *Phys. Rev. B*, **90**, 035423.
24. Jabakhanji, B., Michon, A., Consejo, C., Desrat, W., Portail, M., Tiberj, A., Paillet, M., Zahab, A., Cheynis, F., Lafont, F., Schopfer, F., Poirier, W., Bertran, F., Le Fèvre, P., Taleb-Ibrahimi, A., Kazazis, D., Escoffier, W., Camargo, B. C., Kopelevich, Y., Camassel, J., and Jouault, B. (2014b). Tuning the transport properties of graphene films grown by CVD on SiC(0001): Effect of *in situ* hydrogenation and annealing, *Phys. Rev. B*, **89**, 085422.

25. Janssen, T. J. B. M., Fletcher, N. E., Goebel, R., Williams, J. M., Tzalenchuk, A., Yakimova, R., Kubatkin, S., Lara-Avila, S., and Falko, V. I. (2011a). Graphene, universality of the quantum Hall effect and redefinition of the SI system, *New J. Phys.*, **13**, 093026.

26. Janssen, T. J. B. M., Tzalenchuk, A., Lara-Avila, S., Kubatkin, S., and Fal'ko, V. I. (2013). Quantum resistance metrology using graphene, *Rep. Prog. Phys.*, **76**, 104501.

27. Janssen, T. J. B. M., Tzalenchuk, A., Yakimova, R., Kubatkin, S., Lara-Avila, S., Kopylov, S., and Fal'ko, V. I. (2011b). Anomalously strong pinning of the filling factor $\nu = 2$ in epitaxial graphene, *Phys. Rev. B*, **83**, 233402.

28. Janssen, T. J. B. M., Williams, J. M., Fletcher, N. E., Goebel, R., Tzalenchuk, A., Yakimova, R., Lara-Avila, S., Kubatkin, S., and Fal'ko, V. I. (2012). Precision comparison of the quantum Hall effect in graphene and gallium arsenide, *Metrologia*, **49**, 294.

29. Ji, S.-H., Hannon, J. B., Tromp, R. M., Perebeinos, V., Tersoff, J., and Ross, F. M. (2012). Atomic-scale transport in epitaxial graphene, *Nat. Mater.*, **11**, 2, 114.

30. Jobst, J., Waldmann, D., Gornyi, I. V., Mirlin, A. D., and Weber, H. B. (2012). Electron–electron interaction in the magnetoresistance of graphene, *Phys. Rev. Lett.*, **108**, 106601.

31. Jouault, B., Camara, N., Jabakhanji, B., Caboni, A., Consejo, C., Godignon, P., Maude, D. K., and Camassel, J. (2012). Quantum Hall effect in bottom-gated epitaxial graphene grown on the C-face of SiC, *Appl. Phys. Lett.*, **100**, 052102.

32. Jouault, B., Jabakhanji, B., Camara, N., Desrat, W., Consejo, C., and Camassel, J. (2011). Interplay between interferences and electron–electron interactions in epitaxial graphene, *Phys. Rev. B*, **83**, 195417.

33. Jouault, B., Jabakhanji, B., Camara, N., Desrat, W., Tiberj, A., Huntzinger, J.-R., Consejo, C., Caboni, A., Godignon, P., Kopelevich, Y., and Camassel, J. (2010). Probing the electrical anisotropy of multilayer graphene on the Si-face of 6H-SiC, *Phys. Rev. B*, **82**, 085438.

34. Klitzing, K. V., Dorda, G., and Pepper, M. (1980). New method for high-accuracy determination of the fine-structure constant based on quantized Hall resistance, *Phys. Rev. Lett.*, **45**, 494.

35. Kopylov, S., Tzalenchuk, A., Kubatkin, S., and Fal'ko, V. I. (2010). Charge transfer between epitaxial graphene and silicon carbide, *Appl. Phys. Lett.*, **97**, 112109.

36. Lafont, F., Ribeiro-Palau, R., Kazazis, D., Michon, A., Couturaud, O., Consejo, C., Chassagne, T., Zielinski, M., Portail, M., Jouault, B., Schopfer,

F., and Poirier, W. (2015). Quantum Hall resistance standards from graphene grown by chemical vapour deposition on silicon carbide, *Nat. Commun.* **6**.

37. Lara-Avila, S., Moth-Poulsen, K., Yakimova, R., Bjørnholm, T., Fal'ko, V., Tzalenchuk, A., and Kubatkin, S. (2011). Non-volatile photochemical gating of an epitaxial graphene/polymer heterostructure, *Adv. Mater.*, **23**, 878.
38. Lartsev, A., Yager, T., Bergsten, T., Tzalenchuk, A., Janssen, T. J. B. M., Yakimova, R., Lara-Avila, S., and Kubatkin, S. (2014). Tuning carrier density across Dirac point in epitaxial graphene on SiC by corona discharge, *Appl. Phys. Lett.*, **105**, 063106.
39. Laughlin, R. B. (1981). Quantized Hall conductivity in two dimensions, *Phys. Rev. B*, **23**, 5632.
40. Lien, N. V. (1984). Self-consistent calculations of edge channels in laterally confined two-dimensional electron systems, *Sov. Phys. Semicond.*, **18**, 207.
41. Low, T., Perebeinos, V., Tersoff, J., and Avouris, P. (2012). Deformation and scattering in graphene over substrate steps, *Phys. Rev. Lett.*, **108**, 096601.
42. McCann, E., Kechedzhi, K., Fal'ko, V. I., Suzuura, H., Ando, T., and Altshuler, B. L. (2006). Weak-localization magnetoresistance and valley symmetry in graphene, *Phys. Rev. Lett.*, **97**, 146805.
43. Michon, A., Vézian, S., Roudon, E., Lefebvre, D., Zielinski, M., Chassagne, T., and Portail, M. (2013). Effects of pressure, temperature, and hydrogen during graphene growth on SiC(0001) using propane-hydrogen chemical vapor deposition, *J. Appl. Phys.*, **113**, 203501.
44. Monteverde, M., Ojeda-Aristizabal, C., Weil, R., Bennaceur, K., Ferrier, M., Guéron, S., Glattli, C., Bouchiat, H., Fuchs, J. N., and Maslov, D. L. (2010). Transport and elastic scattering times as probes of the nature of impurity scattering in single-layer and bilayer graphene, *Phys. Rev. Lett.*, **104**, 126801.
45. Morozov, S. V., Novoselov, K. S., Katsnelson, M. I., Schedin, F., Elias, D. C., Jaszczak, J. A., and Geim, A. K. (2008). Giant intrinsic carrier mobilities in graphene and its bilayer, *Phys. Rev. Lett.*, **100**, 016602.
46. Nagayoshi, H., Nakao, K., and Uemura, Y. (1976). Band theory of graphite. I. Formalism of a new method of calculation and the Fermi surface of graphite, *J. Phys. Soc. Jpn*, **41**, pp. 1480–1487.
47. Novoselov, K. S., Geim, A. K., Morozov, S. V., Jiang, D., Katsnelson, M. I., Grigorieva, I. V., Dubonos, S. V., and Firsov, A. A. (2005). Two-

dimensional gas of massless Dirac fermions in graphene, *Nature*, **438**, 7065, 197.

48. Novoselov, K. S., Jiang, Z., Zhang, Y., Morozov, S. V., Stormer, H. L., Zeitler, U., Maan, J. C., Boebinger, G. S., Kim, P., and Geim, A. K. (2007). Room-temperature quantum Hall effect in graphene, *Science*, **315**, 1379.

49. Pallecchi, E., Lafont, F., Cavaliere, V., Schopfer, F., Mailly, D., Poirier, W., and Ouerghi, A. (2014). High electron mobility in epitaxial graphene on 4H-SiC(0001) via post-growth annealing under hydrogen, *Sci. Rep.*, **4**, 4558.

50. Pallecchi, E., Ridene, M., Kazazis, D., Lafont, F., Schopfer, F., Poirier, W., Goerbig, M. O., Mailly, D., and Ouerghi, A. (2013). Insulating to relativistic quantum Hall transition in disordered graphene, *Sci. Rep.*, **3**, 1791.

51. Pan, W., Howell, S. W., Ross, A. J., Ohta, T., and Friedmann, T. A. (2010). Observation of the integer quantum Hall effect in high quality, uniform wafer-scale epitaxial graphene films, *Appl. Phys. Lett.*, **97**, 252101.

52. Perebeinos, V. and Avouris, P. (2010). Inelastic scattering and current saturation in graphene, *Phys. Rev. B*, **81**, 195442.

53. Peres, N. M. R. (2010). *Colloquium*: The transport properties of graphene: An introduction, *Rev. Mod. Phys.*, **82**, 2673.

54. Poirier, W. and Schopfer, F. (2010). Can graphene set new standards? *Nat. Nanotech.*, **5**, 3, 171.

55. Poirier, W. and Schopfer, F. (2009). Resistance metrology based on the quantum Hall effect, *Eur. Phys. J. Special Topics*, **172**, 207.

56. Ponomarenko, L. A., Yang, R., Mohiuddin, T. M., Katsnelson, M. I., Novoselov, K. S., Morozov, S. V., Zhukov, A. A., Schedin, F., Hill, E. W., and Geim, A. K. (2009). Effect of a high-κ environment on charge carrier mobility in graphene, *Phys. Rev. Lett.*, **102**, 206603.

57. Portail, M., Michon, A., Vézian, S., Lefebvre, D., Chenot, S., Roudon, E., Zielinski, M., Chassagne, T., Tiberj, A., Camassel, J., and Cordier, Y. (2012). Growth mode and electric properties of graphene and graphitic phase grown by argon-propane assisted CVD on 3C-SiC/Si and 6H-SiC, *J. Cryst. Growth*, **349**, 27.

58. Prange, R. E. and Girvin, S. M. (Eds) (1990). *The Quantum Hall Effect* (Springer).

59. Qi, Y., Rhim, S. H., Sun, G. F., Weinert, M., and Li, L. (2010). Epitaxial graphene on SiC(0001): More than just honeycombs, *Phys. Rev. Lett.*, **105**, 085502.

60. Riedl, C., Coletti, C., and Starke, U. (2010). Structural and electronic properties of epitaxial graphene on SiC(0001): A review of growth, characterization, transfer doping and hydrogen intercalation, *J. Phys. D Appl. Phys.*, **43**, 374009.

61. Riedl, C., Starke, U., Bernhardt, J., Franke, M., and Heinz, K. (2007). Structural properties of the graphene–SiC(0001) interface as a key for the preparation of homogeneous large-terrace graphene surfaces, *Phys. Rev. B*, **76**, 245406.

62. Satrapinski, A., Novikov, S., and Lebedeva, N. (2013). Precision quantum Hall resistance measurement on epitaxial graphene device in low magnetic field, *Appl. Phys. Lett.*, **103**, 173509.

63. Schopfer, F. and Poirier, W. (2013). Quantum resistance standard accuracy close to the zero-dissipation state, *J. Appl. Phys.*, **114**, 064508.

64. Shklovskii, B. I. and Efros, A. L. (1984). *Electronic Properties of Doped Semiconductors*, Springer-Verlag Berlin Heidelberg.

65. Speck, F., Jobst, J., Fromm, F., Ostler, M., Waldmann, D., Hundhausen, M., Weber, H. B., and Seyller, T. (2011). The quasi-free-standing nature of graphene on H-saturated SiC(0001), *Appl. Phys. Lett.*, **99**, 122106.

66. Strupinski, W., Grodecki, K., Wysmolek, A., Stepniewski, R., Szkopek, T., Gaskell, P. E., Grüneis, A., Haberer, D., Bozek, R., Krupka, J., and Baranowski, J. M. (2011). Graphene epitaxy by chemical vapor deposition on SiC, *Nano Lett.*, **11**, 1786.

67. Tanabe, S., Sekine, Y., Kageshima, H., Nagase, M., and Hibino, H. (2010). Half-integer quantum Hall effect in gate-controlled epitaxial graphene devices, *Appl. Phys. Express*, **3**, 075102.

68. Tanabe, S., Sekine, Y., Kageshima, H., Nagase, M., and Hibino, H. (2011). Carrier transport mechanism in graphene on SiC(0001), *Phys. Rev. B*, **84**, 115458.

69. Tedesco, J. L., VanMil, B. L., Myers-Ward, R. L., McCrate, J. M., Kitt, S. A., Campbell, P. M., Jernigan, G. G., Culbertson, J. C., Eddy, C. R., and Gaskill, D. K. (2009). Hall effect mobility of epitaxial graphene grown on silicon carbide, *Appl. Phys. Lett.*, **95**, 122102.

70. Tikhonenko, F. V., Kozikov, A. A., Savchenko, A. K., and Gorbachev, R. V. (2009). Transition between electron localization and antilocalization in graphene, *Phys. Rev. Lett.*, **103**, 226801.

71. Tzalenchuk, A., Lara-Avila, S., Kalaboukhov, A., Paolillo, S., Syvajarvi, M., Yakimova, R., Kazakova, O., M., J. J. B., Fal'ko, V., and Kubatkin, S. (2010). Towards a quantum resistance standard based on epitaxial graphene, *Nat. Nanotech.*, **5**, 186.

72. Van Bommel, A., Crombeen, J., and Van Tooren, A. (1975). LEED and Auger electron observations of the SiC(0001) surface, *Surf. Sci.*, **48**, 463.

73. Varchon, F., Feng, R., Hass, J., Li, X., Nguyen, B. N., Naud, C., Mallet, P., Veuillen, J. Y., Berger, C., Conrad, E. H., and Magaud, L. (2007). Electronic structure of epitaxial graphene layers on SiC: Effect of the substrate, *Phys. Rev. Lett.*, **99**, 126805.

74. Vitali, L., Schneider, M. A., Kern, K., Wirtz, L., and Rubio, A. (2004). Phonon and plasmon excitation in inelastic electron tunneling spectroscopy of graphite, *Phys. Rev. B*, **69**, 121414.

75. Waldmann, D., Jobst, J., Speck, F., Seyller, T., Krieger, M., and Weber, H. B. (2011). Bottom-gated epitaxial graphene, *Nat. Mater.*, **10**, 5, 357.

76. Wu, X., Hu, Y., Ruan, M., Madiomanana, N. K., Hankinson, J., Sprinkle, M., Berger, C., and de Heer, W. A. (2009). Half integer quantum Hall effect in high mobility single layer epitaxial graphene, *Appl. Phys. Lett.*, **95**, 223108.

77. Zhang, Y., Tan, Y.-W., Stormer, H. L., and Kim, P. (2005). Experimental observation of the quantum Hall effect and Berry's phase in graphene, *Nature*, **438**, 7065, 201.

Appendix A

Raman Spectroscopy of Graphene on Silicon Carbide

Ana Ballestar
Graphene Nanotech, S. L., Miguel Villanueva 3, Logroño, 26001, Spain
ana@gpnt.es

This appendix is intended for scientists and engineers working on epitaxial graphene grown on silicon carbide (SiC); those particularly interested in the characterization of such samples by using Raman spectroscopy.

A.1 Raman as a Tool to Distinguish Graphene

After the isolation of graphene became accessible and its outstanding features had been reported [1, 2], the synthetic production of graphene increased notably. As a matter of fact, the need for an efficient non-invasive tool rose proportionally. Raman spectroscopy appears to be the most adequate and versatile technique available nowadays.

Epitaxial Graphene on Silicon Carbide: Modeling, Devices, and Applications
Edited by Gemma Rius and Philippe Godignon

ISBN 978-981-4774-20-8 (Hardcover), 978-1-315-18614-6 (eBook)
www.panstanford.com

A.1.1 Raman Scattering

The Raman spectroscopy method is based on an inelastic scattering of light phenomenon, known as Raman scattering. This process involves the excitation of electrons from a ground state by a monochromatic light (usually a laser), which after hitting a molecule experience an energy relaxation driven by lattice vibrations.

After relaxation, the energy of some of the photons is changed and the difference between the incident and the scattered light energy is measurable. This difference is the so-called Raman shift (see Fig. A.1). Even though the ratio of Raman-scattered light to total scattered light is small, Raman scattering spectroscopy is a powerful tool for investigations on lattice vibrations and electron states. The specific nature of these properties makes Raman spectra a fingerprint for each material.

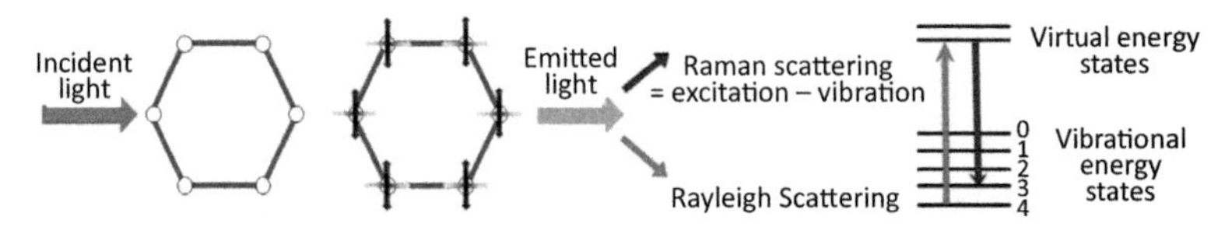

Figure A.1 Schematic representation of the Raman scattering mechanism. The incident light induces a vibration of the molecule, and the electrons around the bonds experience a momentary distortion. After relaxation, some photons of the emitted light change their frequency (Raman scattered). The energy difference between the ground state and the new state is the Raman shift, i.e., $E_4 - E_3$, following the notation on the sketch on the right-hand side.

A.1.2 Raman Spectroscopy

A Raman spectrometer setup consists on a light source, a monochromator, a sample holder, and a detector. Among the different techniques existing, confocal Raman microscopy, which basically combines an optical microscope with the Raman spectrometer, is the most interesting option for graphene investigations. Here are some of its advantages:

- Reduction of the laser spot, which allows analysis of features with dimensions ~350 nm.
- The detectors only collect Raman signals, increasing resolution.

- Possibility of recording Raman mappings, providing information about homogeneity of the sample.
- Depth spatial resolution, allowing for analysis of multilayered samples.

Raman spectroscopy is also non-invasive and fast; samples do not require specific preparation. It provides high resolution and specific information for identifying as well as evaluating materials, such as determining their crystal quality, and it is potentially applicable to industrial-scale production.

A.1.3 Raman Spectroscopy of Graphene

The chemical sensitivity of Raman spectroscopy allows the investigation of the vibrational modes related to the sp^2 and sp^3 carbon bonds and directly measures the phonon dispersion bands related to the A-B carbon atoms, which define the hexagonal lattice of graphene. Due to the graphene band structure, an incident light connects two real electronic states (it usually connects a virtual energy state with the ground state, see Fig. A.1), which results in a very strong phonon signal in the spectra. Three prominent features are the fingerprint of graphene Raman spectrum: the so-called G, D, and 2D bands, which lie in ~1580 cm^{-1}, ~1350 cm^{-1}, and ~2700 cm^{-1}, respectively.

The G band is an in-plane vibrational mode involving the sp^2-hybridized carbon atoms of the graphene sheet. The position of the G peak (ω_G) and the full width at half-maximum (FWHM, Γ_G) can be correlated to the number of layers: ω_G shifts to lower energy and Γ_G increases as the layer thickness increases [3, 4]. However, this must be carefully applied as ω_G can be strongly affected by doping [5, 6], strain [7], and temperature.

The D band is due to the breathing mode of sp^2 atoms in a carbon ring adjacent to an edge or a defect. Therefore, this band is typically used to evaluate the presence of short-range disorder of graphene samples, as its intensity is directly proportional to the amount of defects.

The 2D band is the second order scattering process of the D line resulting from a two phonon lattice vibrational process. Unlike the D band, it does not require to be activated by the proximity of a defect. Hence, the 2D mode is always present even in defect-free graphene samples. The changes in the electronic structure due to an increase in the number of layers are captured in changes of its shape, width (Γ_{2D}) and position of the 2D peak (ω_{2D}) [3]. Therefore, it serves to distinguish mono-, bi-, and few-layer graphene. For single-layer graphene, the 2D peak is symmetric and can be fitted to a single Lorentzian with Γ_{2D} of ~30 cm^{-1}. Adding successive layers causes the lowering of the symmetry, which induces 2D band splitting into several overlapping modes. Consequently, the 2D peak for bilayer or few-layer graphene requires four or more Lorentzians for an adequate fit. Moreover, the 2D band is known to be doping dependent [8]. If the charge density n increases, ω_{2D} increases for p-doping and decreases for n-doping [9].

Finally, the analysis of the ratio of the relative integrated intensities of the 2D and G peak (I_{2D}/I_G) identifies high-quality (defect-free) graphene and informs about the charge density.

Considering the advantages Raman spectroscopy provides for the investigation of graphene, it can be considered the indispensable and effective tool that the nascent field of carbon nanomaterials demands.

A.2 Raman Spectra of Graphene on Silicon Carbide

Even though the line shapes of the spectra measured on epitaxial graphene on SiC agree with the ones obtained for graphene on other substrates (e.g., exfoliated, freestanding), some differences are observed and, therefore, hereby discussed.

The main disagreement is that all the Raman bands of epitaxial graphene are typically blue-shifted with respect to those of exfoliated samples, revealing a phonon hardening effect [10]. The origin of the significant shift (~17 cm^{-1} for the G band and ~42 cm^{-1} for the 2D band [11]) resides in the compressive strain present in graphene grown on the Si-face of an SiC substrate. The mismatch between graphene and the substrate induces a biaxial stress calculated in ~2.27 GPa [11]. The red-shift of the Raman peaks observed in epitaxial suspended graphene, as compared to epitaxial graphene on SiC [12], points out the vanishing of the strain after suspension of graphene. The contribution to the blue-shift resulting from electron and/or hole doping is small, as doping concentrations larger than 1.5×10^{13} are required to induce such shifts in ω_G [5]. The weaker dependence of the 2D shift on doping (10–30% less as compared with the G band dependence [13]) also points to the strain effect as responsible for the large 2D band shift. Blue-shift is mainly observed on epitaxial graphene grown on the Si-face of SiC wafers.

Contrarily, note that if graphene is grown on the C-face of SiC, almost no strain at the graphene layer to substrate interface appears; consequently, the Raman signature is similar to exfoliated graphene samples [14]. However, other major disadvantages occur, i.e., more inhomogeneities in thickness and the stacking order is much less trivial to be analyzed [15].

A.2.1 Substrate Contribution

The Raman spectra of epitaxial graphene on SiC (0001) exhibit additional peaks between 1450 and 1750 cm^{-1} due to the intrinsic phonon modes of the SiC substrate. These features are superimposed upon the G band of graphene, preventing an accurate analysis (see Fig. A.2a). For precise graphene analysis, one has to subtract the spectra of pristine SiC. The resulting spectra will readily show the characteristic D, G, and 2D peaks (see Fig. A.2b). As described previously, the peaks' parameters provide the necessary information for determination of the number of layers, strain, etc. However, some of the outlined rules for exfoliated graphene samples are not strictly applicable to epitaxial graphene.

Furthermore, the intrinsic interface structure between epitaxial graphene and SiC, consisting of a graphene-like lattice of atoms covalently bonded to some of the underlying silicon atoms, the so-called buffer layer, introduces non-vanishing signals in the Raman spectra. Two additional broad features at ~1355 and ~1580 cm^{-1} appear [17] (see Fig. A.3). Although small signs of these features remain in the line shape of epitaxial graphene after subtraction of bare SiC spectra, the clearness of the characteristic peaks is sufficient for proper analysis (see Fig. A.2b).

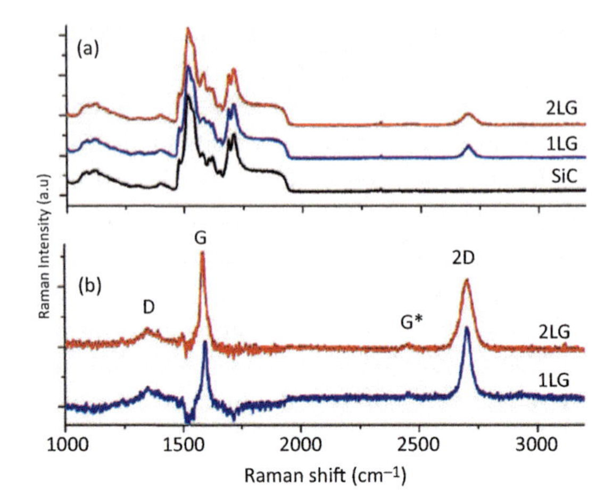

Figure A.2 Typical Raman spectra of epitaxial graphene on SiC(0001) for single layer (blue), bilayer (red), as well as the spectra of a bare SiC substrate (black): (a) as measured, (b) after subtraction of substrate contribution [16].

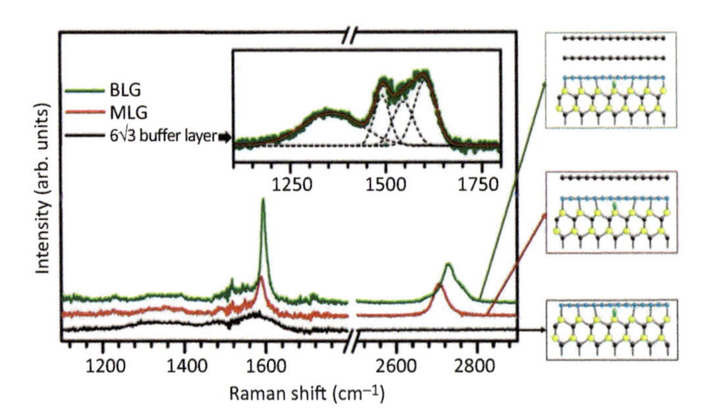

Figure A.3 Raman spectra of epitaxial graphene on SiC(0001) after subtraction of SiC Raman spectra for single-layer graphene, MLG (red), bilayer graphene, BLG (green), and buffer layer (black). The inset shows in detail the characteristic features of the buffer layer. On the right-hand side, the schematic crystalline cross sections of BLG, MLG, and buffer layer, from top to bottom, are shown. The yellow circles correspond to Si atoms and the black ones to C, both forming the SiC substrate structure. On top of them, the line made of blue dots represents the buffer layer. Note that not every Si atom is bonded to the buffer layer, as indicated with a green drop-shaped symbol. Subsequently, the layers above made of black circles correspond to one or two layers of graphene, for respectively, monolayer graphene and bilayer graphene [17].

For the sake of certainty, the G band must be carefully analyzed. The absence of extra features around the 2D band frequency makes it a more direct tool for the analysis of graphene Raman spectra. Alternatively, there exist different methods to only enhance the graphene signal, while the SiC spectrum is suppressed, e.g., deposition of silver nanoparticles on graphene [18]. Ferralis et al. used a depolarized Raman spectrum, which enhances the intensity of the G band with respect to the SiC features [19].

If graphene is grown on the C-face of SiC, the bonding and the interaction with the substrate are weaker and almost negligible. Hence the measured spectra present none of the just mentioned issues.

A.2.2 The D Band

The D band appears in the Raman spectra of epitaxial graphene at the same frequency it does for exfoliated or suspended samples, i.e., $\sim$1350 cm^{-1}. It indicates the presence of defects, which may arise from corrugation, interactions with the substrate and vacancies, or surface dislocations [11], etc. The larger the intensity of the D peak, the larger the amount of defects on the sample. For example, the intensity of the D band can eventually increase after lithography and/or etching, as disorder and contamination can be introduced in both processes [20].

A.2.3 The G Band

The G band falls between 1580 and 1590 cm^{-1} and exhibits a single Lorentzian shape with FWHM of $\sim$15 cm^{-1} in the Raman spectra of epitaxial graphene. Neither its position (ω_G), nor its width (Γ_G), varies with the number of layers; however, they strongly depend on strain and doping. The G peak positon ω_G indicates strong compressive strain when shifts to higher frequencies [11]. Doping also participates in the shift of the G band phonons, although its contribution is smaller [10]. Nevertheless, the position of the G peak is a sensitive parameter for the estimation of doping: ω_G increases if doping with electrons, and decreases if doping with holes, up to a limit in charge density (n) [5, 6]. The doping level can be determined from ω_G and Γ_G using explicit formulas as done in Ref. [6].

A.2.4 The 2D Band

Despite the fact that the 2D band is sensitive to strain (same behavior as the G band), and doping (shifts to higher frequencies for p-doping and to lower frequencies for n-doping, with increasing n [9]), the major effects on its position (ω_{2D}), width (Γ_{2D}), and shape come from the number of layers and their stacking order. Hence, the characteristics of the 2D peak are widely used to determine the number of layers of graphene samples.

The values of ω_{2D} range between 2700 and 2770 cm^{-1}. It is generally accepted that the 2D band of epitaxial graphene is broader than that of exfoliated samples, probably related to poorer crystallinity. It seems to be a good criterion to assume that a symmetric single Lorentzian peak centered at ~2740 cm^{-1} and Γ_{2D} below 60 cm^{-1} corresponds to the presence of epitaxial single-layer graphene on SiC.

If the number of layers increases, Γ_{2D} exceeds 60 cm^{-1}. This broadening and its multi-peak structure come from the splitting of the band structure at the Dirac points. Hence, by investigating the shape of the 2D band, the precise number of layers can be quantified (as done in exfoliated samples [3]), as well as different couplings between Bernal parts in the stacking sequence can be identified [15]. For instance, non-symmetric 2D peaks at ~2730 cm^{-1}, which are fitted using four Lorentzians, correlate to the presence of bilayers, as done in Ref. [10] (see also Fig. A.2). However, occasionally the 2D band for bilayer and trilayer samples is best fitted with one single Lorentzian (Γ_{2D} of ~65–75 cm^{-1}) and the only difference between them comes from the peak position ω_{2D}. Different interpretations arise here: In some cases, it is believed that an increase in the number of layers derives into a decrease in ω_{2D} [10]; in some others, ω_{2D} increases with the number of layers [7]. The first option relies on the fact that since the thickness increases, the interaction with the substrate becomes weaker and the strain between graphene and substrate relaxes. The second one is possibly justified in terms of the different electronic properties of samples with different thicknesses.

Another criterion would be to simply use the values of Γ_{2D}. As already mentioned, widths below 60 cm^{-1} are generally related to single-layer samples. For graphene grown on SiC(0001), Γ_{2D} exhibits a linear relationship with the inverse number of layers (N), which follows: $\Gamma_{2D} = (-45(1/N) + 88)$ [7]. Note that in the case of samples grown on the C-face of SiC, the width of the 2D band does not change with thickness. Instead, the intensity of the 2D increases notably and the line shape of the Raman spectrum is extremely similar to that of single-layer graphene [15]. Attention must be paid to avoid misleading interpretations of such cases.

A.2.5 Relative Intensity

As done for exfoliated samples, the ratio of integrated intensities of the 2D peak and the G peak, I_{2D}/I_G, is used to investigate physical properties in epitaxial samples. The usual value of I_{2D}/I_G in this case is ~1. However, different values can be found in the literature and are interpreted as an indicator for doping or charge densities n. To be precise, it has been found that I_{2D}/I_G decreases for increasing n [9]. For example, a value of $I_{2D}/I_G = 6$ indicates a weak residual doping of $\sim 10^{12}$ cm^{-2} [14].

A.2.6 SiC Morphology Reflected on Raman Mappings

The stepped morphology of epitaxial graphene grown on the Si-face of SiC is reflected on the so-called Raman mappings. A 2D Raman mapping is generated by performing spectroscopy at each point and extracting quantities, such as peak position or FWHM of the selected peaks. The mappings of epitaxial graphene follow the topography as seen, for instance, under an optical microscope (see Fig. A.4). The growth of epitaxial graphene typically combines the presence of single-layer and bilayer in regions with sizes of the order or below micrometers. Hence, it is extremely important to minimize the laser spot size and control the lateral resolution, so that mixed contributions from differently covered areas can be avoided.

The typical distribution of epitaxial graphene on SiC (0001), as reported previously [21, 22], is shown in Fig. A.4b. On the terrace region, a relatively good uniformity in $\omega_G \sim 1625$ cm^{-1} (not shown in the picture for clarity) and Γ_{2D} of ~30 cm^{-1} is shown, indicating the existence of single-layer graphene. On the step edges, Γ_{2D} broadens up to ~72 cm^{-1} and the Raman peak can be fitted with four Lorentzians (see Fig. A.4c), indicating the presence of bilayer graphene. Investigations of the relative positions of the G and 2D bands revealed that at the step edges, the electron concentration is lower than that on the terraces [21].

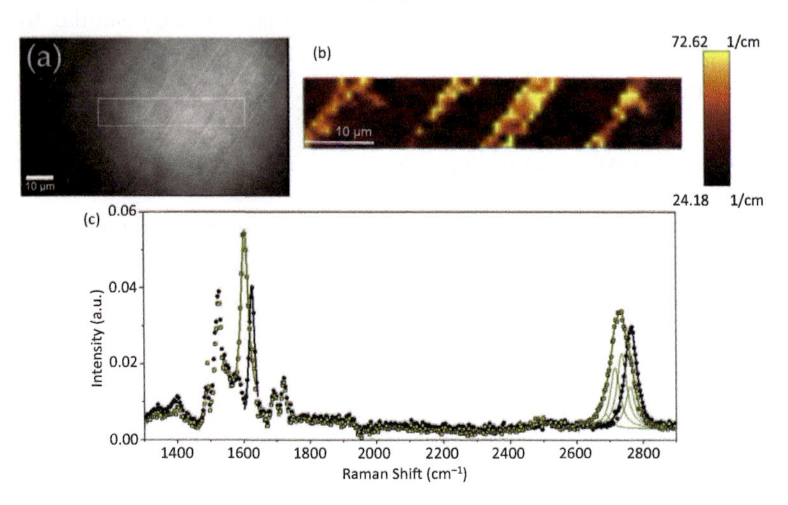

Figure A.4 (a) Optical image of epitaxial graphene on SiC(0001). (b) Raman mapping of the Γ_{2D} from the area inside the white box in (a), and the corresponding scale: yellow areas represent Γ_{2D} ~72 cm^{-1} and dark areas Γ_{2D} ~27 cm^{-1}. (c) Raman spectra representative of the bright (yellow squares) and dark (black dots) areas, respectively. These results indicate that terraces are covered with single-layer graphene, and bilayer graphene is found on the step edges. Results obtained from a sample provided by Graphene Nanotech, S.L.

A.3 Concluding Remarks

The major advantages of Raman spectroscopy for the characterization of epitaxial graphene have been reviewed throughout this chapter. The use of the characteristic parameters ω_{2D}, Γ_{2D}, ω_G, Γ_G, I_D, I_{2D}/I_G has been demonstrated to serve as a tool for informing about the number of layers, strain, and doping of graphene grown from and on SiC substrates. However, for the sake of completeness, complementary techniques, e.g., electron transmission microscopy, transport measurements, and atomic force microscopy such as Kelvin probe force microscopy mode or XPS are highly recommended.

References

1. Novoselov, K. S., Geim, A. K., Morozov, S. V., Jiang, D., Zang, Y., Dubonos, S. V., Grigorieva, I. V., and Firsov, A. A. (2004). Electric field effect in atomically thin carbon films, *Science*, **306**, pp. 666–669.

2. Novoselov, K. S., Fal'ko, V. I., Colombo, L., Gellert, P. R., Schwab, M. G., and Kim, K. (2012). A roadmap for graphene, *Nature*, **490**, pp. 192–200.

3. Ferrari, A. C., Meyer, J. C., Scardaci, V., Casiraghi, C., Larezzi, M., Mauri, F., Piscanec, S., Jiang, D., Novoselov, K. S., Roth, S., and Geim, A. K. (2006). Raman spectrum of graphene and graphene layers, *Phys. Rev. Lett.*, **97**, 187401.

4. Wang, H., Wang, Y., Cao, X., Feng, M., and Lan, G. (2009). Vibrational properties of graphene and graphene layers, *J. Raman Spectrosc.*, **40**, pp. 1791–1796.

5. Das, A., Pisana, S., Chakraborty, B., Piscanec, S., Saha, S. K., Waghmare, U. V., Novoselov, K. S., Krishnamurthy, H. R., Geim, A. K., Ferrari, A. C., and Sood, A. K. (2008). Monitoring dopants by Raman scattering in an electrochemically top-gated graphene transistor, *Nat. Nanotechnol.*, **3**, pp. 210–215.

6. Lazzeri, M. and Mauri, F. (2006). Nonadiabatic Khon anomaly in a doped graphene monolayer, *Phys. Rev. Lett.*, **97**, 266407.

7. Lee, D. S., Riedl, C., Krauβ, B., von Klitzing, K., Starke, U., and Smet, J. H. (2008). Raman spectra of epitaxial graphene on SiC and of epitaxial graphene transferred to SiO_2, *Nano Lett.*, **8**, pp. 4320–4325.

8. Basko, D. M., Piscanec, S., and Ferrari, A. C. (2009). Electron–electron interactions and doping dependence of the two-phonon Raman intensity in graphene, *Phys. Rev. B*, **80**, 165413.

9. Caridad, J. M., Rossella, F., Bellani, V., Maicas, M., Patrini, M., and Díez, E. (2010). Effects of particle contamination and substrate interaction on the Raman response of unintentionally doped graphene, *J. Appl. Phys.*, **108**, 084321.

10. Röhl, J., Hundhausen, M., Emtsev, K. V., Seyller, Th., Graupner, R., and Ley, L. (2008). Raman spectra of epitaxial graphene on SiC (0001), *Appl. Phys. Lett.*, **92**, 201918.

11. Ni, Z. H., Chen, W., Fan, X. F., Kuo, J. L., Yu, T., Wee, A. T. S., and Shen, Z. X. (2008). Raman spectroscopy of epitaxial graphene on a SiC substrate, *Phys. Rev. B*, **77**, 115416.

12. Shivaraman, S., Jobst, J., Waldmann, D., Weber, H. B., and Spencer, M. G. (2013). Raman spectroscopy and electrical transport studies of free-standing epitaxial graphene: Evidence of an AB-stacked bilayer, *Phys. Rev. B*, **87**, 195425.

13. Yan, J., Zhang, Y. B., Kim, P., and Pinczuk, A. (2007). Electric field effect of electron–phonon coupling in graphene, *Phys. Rev. Lett.*, **98**, 166802.

14. Camara, N., Jouault, B., Caboni, A., Jabakhanji, B., Desrat, W., Pausas, E., Consejo, C., Mestres, N., Godignon, P., and Camassei, J. (2010). Growth of monolayer graphene on 8º off-axis 4H-SiC (000-1) substrates with application to quantum devices, *Appl. Phys. Lett.*, **97**, 093107.

15. Camara, N., Tiberj, A., Jouault, B., Caboni, A., Jabakhanji, B., Mestres, N., Godignon, P., and Camassel, J. (2010). Current status of self-organized epitaxial graphene ribbons on the C face of 6H-SiC substrates, *J. Phys. D Appl. Phys.*, **43**, 374011.

16. Kazakova, O., Panchal, V., and Burnett, T. L. (2013). Epitaxial graphene and graphene-based devices studied by electrical scanning probe microscopy, *Crystals*, **3**, pp. 191–233.

17. Fromm, F., Oliveira Jr., M. H., Molina-Sánchez, A., Hundhausen, M., Lopes, J. M. J., Riechert, H., Wirtz, L., and Seyller, T. (2013). Contribution of the buffer layer to the Raman spectrum of epitaxial graphene on SiC(0001), *New J. Phys.*, **15**, 043031.

18. Sekine, Y., Hibino, H., Oguri, K., Akazaki, T., Kageshima, H., Nagase, M., Sasaki, K., and Yamaguchi, H. (2013). Surface-enhancement Raman scattering of graphene on SiC, *NTT Tech. Rev.*, **8**, pp. 1–6.

19. Ferralis, N., Maboudian, R., and Carraro, C. (2008). Evidence of structural strain in epitaxial graphene layers on 6H-SiC (0001), *Phys. Rev. Lett.*, **101**, 156801.

20. Shivaraman, S., Barton, R. A., Yu, X., Alden, J., Herman, L., Chandrashekhar, M. V. S., Park, J., McEuen, P. L., Parpia, J. M., Harold, G. C., and Spencer, M. G. (2009). Free-standing epitaxial graphene, *Nano Lett.*, **9**, pp. 3100–3105.

21. Grodecki, K., Bozek, R., Strupinsky, W., Wysmolek, A., Stepniewski, R., and Baranowski, J. M. (2012). Micro-Raman spectroscopy of graphene grown on stepped 4H-SiC (0001) surface, *Appl. Phys. Lett.*, **100**, 261604.

22. Oliveira, M. H., Schumann, T., Ramsteiner, M., Lopes, J. M., and Richert, H. (2011). Influence of the silicon carbide surface morphology in the epitaxial graphene formation, *App. Phys. Lett.*, **99**, 111901.

Appendix B

Graphene on SiC: Chemico-Physical Characterization by XPS

Micaela Castellino[a] **and Jordi Fraxedas**[b]
[a]*Center for Space Human Robotics, Istituto Italiano di Tecnologia, Corso Trento 21, Torino, 10129, Italy*
[b]*Catalan Institute of Nanoscience and Nanotechnology (ICN2), CSIC and The Barcelona Institute of Science and Technology, Campus UAB, Bellaterra, 08193, Barcelona, Spain*
micaela.castellino@iit.it

In this appendix, an overview of the fundamentals of X-ray photoelectron spectroscopy (XPS) analysis will be given and applied to the characterization of epitaxial graphene. Selected samples of graphene grown on SiC substrates will be proposed to exemplify the information that such well-established analysis method can provide.

B.1 XPS: A Powerful Technique for Surface Analysis

The XPS technique is ideally suited for the investigation of the electronic structure of solid surfaces. The term electronic structure

Epitaxial Graphene on Silicon Carbide: Modeling, Devices, and Applications
Edited by Gemma Rius and Philippe Godignon

ISBN 978-981-4774-20-8 (Hardcover), 978-1-315-18614-6 (eBook)
www.panstanford.com

includes both the determination of *semi-quantitative atomic concentration* as well as the *chemical state* of the involved elements. Such (local) probe analysis can be extended to the (i) generation of images (the so-called *mapping mode*), (ii) in-plane microscopy, to the variation in chemistry across the surface either in conjunction with ion sputtering or by using hard X-rays, (iii) structural determination using electron diffraction, and even can be applied to (iv) liquid surfaces.

The surface sensitivity of XPS, typically below 3 nm in depth, arises from the short mean free path of electron in solids. Since the mean free path depends on the kinetic energy of the out-coming electrons, experiments can be designed in order to characterize only the most external surface layer, which is the ideal experimental situation for the case of graphene. A detailed description about XPS applied to carbon-based materials can be found in Ref. [1].

B.1.1 Planar Approach Analysis

A typical XPS measurement starts with the so-called *survey* or *wide-range* spectrum, in which the entire available energy range of the X-ray source is checked. With this analysis, we can evaluate the presence of different chemical elements and calculate their relative atomic concentrations. After this preliminary and mandatory step, we can focus our attention on the main photoemission peaks, in order to perform a deeper and finer analysis, which can be called a *high resolution* (HR) investigation.

In our case studies, (epitaxial) graphene on SiC, we are interested mainly in the C1s and possibly in the Si2p peaks, which can give valuable information regarding the deposited graphene material and its underlying substrate. Moreover, we can also explore the O1s peak, which can be related to contamination, especially for ex situ grown samples, which would have been inevitably exposed to the atmosphere. A typical layout for the C1s peak is shown in Fig. B.1 [2]. The line is composed of two distinctive features: a smaller peak for binding energies (E_B) lower than 284.0 eV and a more prominent one for E_B higher than 284.0 eV. The first one is associated to the chemical shift due to the C atom involved in a carbide compound and arises from the SiC substrate. The second one is a convolution of several peaks: The one located at 284.5 eV is attributed to the sp^2 allotropic form of C, which is the one we expect from the graphene layers. The second component at 285.0 eV is due to the sp^3 diamond-like allotropic form, and it takes origin mainly from C–H bonding and eventually from defects that are present in the material. The remaining peaks above 285.5 eV are caused by carbon–oxygen bonding such as C–O (286.6 eV), C=O (287.8 eV), and O=C–O (289.0 eV). It has to be highlighted that the attribution of the C1s chemical shifts is sometimes quite tricky and cumbersome, since a large dispersion in energies can be actually found [3, 4]. Such discrepancies mainly arise from the selection of the energy reference in the case of insulating materials. The C1s is typically assigned to 284.8 eV for adventitious carbon, but such an artificial procedure can falsify influence the peak assignment of the different species [5–8]. The correct selection of the real energy reference is an extremely important point in any photoemission characterization.

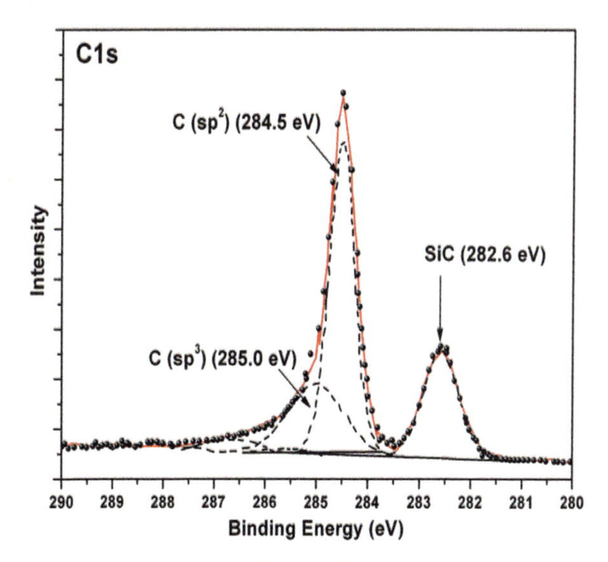

Figure B.1 XPS C1s peak of a thin (5 layers) graphene film grown on an SiC substrate. Black dots are experimental data, dotted black lines are deconvolution curves, and red line is the final fit curve. Reprinted from Ref. [2], with permission from Elsevier.

Once we have completed the (spectrum) data processing by using the peak deconvolution procedure, we can also evaluate the relative percentage for each kind of bond. In the example, it shows 24.6% for C in SiC bonds, 44.6% for sp^2, 21.0% for sp^3, and the residual to oxidized forms. As we can see from the previous results, we have quite a considerable contribution from the C atoms present in the SiC substrate. If we wish to reduce this contribution, we can change the measurement geometry, to be much more sensitive to the real surface of our sample, as follows.

B.1.2 Angle-Resolved XPS (AR-XPS)

The probing depth can be varied either by changing the collection angle or by tuning the electron kinetic energy with synchrotron radiation. While the first method can be performed in the laboratory using conventional X-ray sources with a single photon energy, the second one requires the use of synchrotron radiation sources. Here, we shortly comment on the first, more common and affordable approximation.

In angle-resolved XPS (AR-XPS), the surface sensitivity changes as a function of the collection angle. Electrons acquired at a grazing (*surface*) angle come exclusively from a shallow region of the sample. Electrons acquired at a near-normal (*bulk*) angle come from a deeper region from the sample (see Fig. B.2, left). Spectra acquired from thin films on substrates are thus affected by the collection angle. In the here proposed example (see Fig. B.2, right), a thin graphene layer has been analyzed at two different angles: at 90° (red dots curve) and at 20° (black dots curve). From the bulk angle spectrum, both SiC and graphene show similar intensities. On the surface angle spectrum, the intensity of SiC is much lower as compared to the graphene signal, since the C1s component at nearly 283.0 eV has been strongly reduced compared to the one at 284.5 eV. As the ratio of SiC to graphene is much lower on the surface angle, this fact points out that graphene is present, predominantly, on the surface, with SiC as the substrate. Moreover, in the surface angle spectrum, the graphene peak is much more asymmetric, showing an increase in the intensity of the C–O region. Thus, some oxygen is present in the first layer. Commercial software can provide a relative depth plot from AR-XPS data. The model reconstructed for this example gives a graphene thickness of 0.81 nm (Courtesy of Thermo Fisher Scientific). The surface thickness (d) can also be calculated by means of the two-layer model [9]:

$$d = \lambda_{\text{sub}} \cdot \sin\theta \cdot \ln\left(\frac{1}{\beta} \cdot \frac{I_{\text{surf}}}{I_{\text{sub}}} + 1\right) \tag{B.1}$$

where λ_{sub} is the attenuation length of photoelectrons from the substrate SiC region traveling through the graphene, θ is the angle between the sample surface plane and the electron analyzer, β is a parameter dependent of the surface and bulk composition respectively, and $I_{\text{surf}}/I_{\text{sub}}$ is the ratio of intensities of the surface film and bulk substrate, respectively [10]. The AR-XPS would give the same kind of information if obtained by a simple depth profile XPS, but the latter mode is destructive, since the sputtering process with ions irreversibly modifies (removal) the surface material.

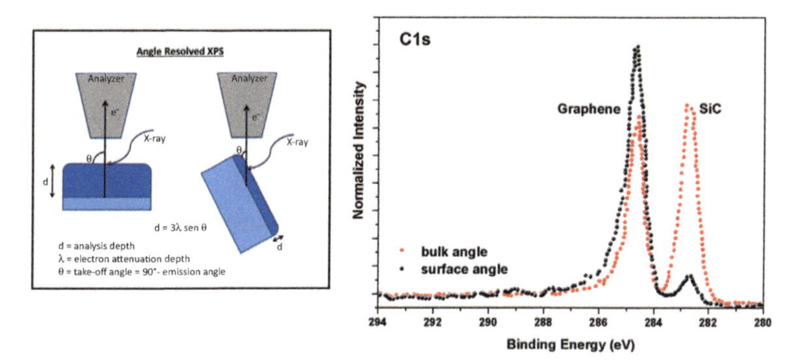

Figure B.2 (Left) Scheme of the planar versus AR-XPS acquisition mode. (Right) XPS C1s peaks for an epitaxial graphene material taken at 90° (bulk angle, red spots) and 20° (surface angle, black spots).

B.1.3 Mapping Mode

Nowadays, many XPS equipment are equipped with position-sensitive detectors (PSD), such as multichannel plate (MCP). With such devices, it is possible to perform HR acquisition in a chemical mapping mode, which will give an element-specific imaging of the sample surface. For example, if we want to assess the chemical modification of ultra-thin graphene layers or explore the layer uniformity, we can use mapping to collect XPS data across the film surface and be able to identify the present chemical states. We have to perform HR scans for each element and/or for each chemical shift we want to check. At the end of the survey series, we can get information regarding the spatial distribution of chemical elements on top of the substrate, such as the presence of areas either covered by continuous graphene film or scattered islands of graphene with different thickness all over the substrate [11]. This kind of analysis is time consuming depending on the size and number of points that have to be acquired, but indeed provides valuable information on the homogeneity of the surface.

B.1.4 Using Synchrotron Radiation

Dedicated electron storage rings provide an intense, continuous, and highly collimated spectrum of light (*white light*), known as synchrotron radiation. Monochromators permit the selection of target photon energies, so that when using photoemission spectroscopy, the kinetic energy of the emitted photoelectron can be precisely chosen. In this way, the surface sensitivity can be tuned. The highest surface sensitivity is achieved for kinetic energies of about 100 eV. Thus, in the case of the C1s line, photon energies of about 390 eV must be selected. Several works of XPS using synchrotron radiation can be found in the literature (see, e.g., Ref. [12]). Remarkably, with such studies, the atomic arrangement of the different layers can be exactly elucidated: from pristine graphene-like arrangement (negligible graphene–SiC surface interaction) on top to covalent bonding with the SiC substrate in the sub-surface (buffer-layer effect).

References

1. Castellino, M., Rius, G., Virga, A., and Tagliaferro, A. (2016). Nanographene patterns from focused ion beam-induced deposition structural characterization of graphene materials by XPS and raman scattering, in *Graphene Science Handbook*, Aliofkhazraei, M., Ali, N., Milne, W.I., Ozkan, C. S., Mitura, S., Gervasoni, and J. L. (Eds.), CRC Press.
2. Giusca, C. E., Spencer, S. J., Shard, A. G., Yakimova, R., and Kazabova, O. (2014). Exploring graphene formation on the C-terminated face of SiC by structural, chemical and electrical methods, *Carbon*, **69**, pp. 221–229.

3. Castellino, M., Stolojan, V., Virga, A. Rovere, M., Cabiale, K., Galloni, M. R., and Tagliaferro, A. (2013). Chemico-physical characterisation and in vivo biocompatibility assessment of DLC-coated coronary stents, *Anal. Bioanal. Chem.*, **405**, pp. 321–329.
4. Yannopoulos, S. N., Siokou, A., Nasikas, N. K., Dracopoulos, V., Ravani, F., and Papatheodorou, G. N. (2012). CO_2-laser-induced growth of epitaxial graphene on 6H-SiC(0001), *Adv. Func. Mater.*, **22**, pp. 113–120.
5. Glenn, G. J., van Mil, B. L., Tedesco, J. L., Tischler, J. G., Glaser, E. R., Davidson, A., Campbell, P. M., and Gaskill, D. K. (2009). Comparison of epitaxial graphene on Si-face and C-face 4H-SiC formed by ultrahigh vacuum and RF furnace production, *Nano Lett.*, **9**, pp. 2605–2609.
6. Riedl, C., Coletti, C., and Starke, U. (2010). Structural and electronic properties of epitaxial graphene on SiC(0001): A review of growth, characterization, transfer doping and hydrogen intercalation, *J. Phys. D Appl. Phys.*, **43**, 374009.
7. Park, J., Mitchell, W. C., Grazulis, L., Smith, H. E., Eyink, K. G., Boeckl, J. J., Tomich, D. H., Pacley, S. D., and Hoelscher, J. E. (2010). Epitaxial graphene growth by carbon molecular beam epitaxy (CMBE), *Adv. Mater.*, **22**, pp. 4140–4145.
8. Pierucci, D., Sediri, H., Hajlaoui, M., Girard, J. C., Brumme, T., Calandra, M., Velez-Fort, E., Patriarche, G., Silly, M. G., Ferro, G., Soulière, V., Marangolo, M., Sirotti, F., Mauri, F., and Ouerghi, A. (2015). Evidence for flat bands near the Fermi level in epitaxial rhombohedral multilayer graphene, *ACS Nano*, **9**, pp. 5432–5439.
9. Hill, J. M., Royce, D. G., Fadley, C. S., Wagner, L. F., and Grunthaner, F. J. (1976). Properties of oxidized silicon as determined by angular-dependent X-ray photoelectron spectroscopy, *Chem. Phys. Lett.*, **44**, pp. 225–231.
10. Shivaraman, S., Chandrashekhar, M. V. S., Boeckl, J. J., and Spencer, M. G. (2009). Thickness estimation of epitaxial graphene on SiC using attenuation of substrate Raman intensity, *J. Elec. Mater.*, **38**, pp. 725–730.
11. http://xpssimplified.com/imaging_mapping.php
12. Emtsev, K. V., Speck, F., Seyller, Th., Ley, L., and Riley, J. D. (2008). Interaction, growth, and ordering of epitaxial graphene on SiC{0001} surfaces: A comparative photoelectron spectroscopy study, *Phys. Rev. B*, **77**, 155303.

Index

PGSTL 05/01/2018